YOUR AMAZING BODY

Your Amazing Body

Steve Laufmann

Howard Glicksman

Seattle   Discovery Institute Press   2025

Description

Every day your body must solve hundreds of hard engineering problems simultaneously, or else you'll die. Your body enables you to see and hear. It heals you from injuries and diseases. It maintains your temperature. It makes it possible for you to wield a sledgehammer during the day and play piano sonatas at night. Every minute your body choreographs a host of complex systems that are required for you to live.

Join engineer Steve Laufmann and physician Howard Glicksman as they take you on an incredible journey exploring some of your body's greatest marvels, including how your hearing and vision work, how you coordinate your movements, and—perhaps the greatest miracle of them all—how you developed from a single cell at conception. Along the way, you will discover compelling evidence that our bodies are the handiwork of a master designer and engineer.

Copyright Notice

Library Cataloging Data

Your Amazing Body by Steve Laufmann and Howard Glicksman

Cover design by Tri Widyatmaka.

154 pages, 6 x 9 inches

Library of Congress Control Number: 2025935848

ISBN: 978-1-63712-077-4 (paperback), 978-1-63712-078-1 (Kindle), 978-1-63712-079-8 (EPUB)

BISAC: SCI036000 SCIENCE/
    Life Sciences/Human Anatomy & Physiology

BISAC: SCI027000 SCIENCE/
    Life Sciences/Evolution

BISAC: TEC073000 TECHNOLOGY & ENGINEERING/
    Systems Engineering

Publisher Information

Discovery Institute Press, 208 Columbia Street, Seattle, WA 98104

Internet: discovery.press

Published in the United States of America on acid-free paper.

First Edition, August 2025

Praise for
Laufmann and Glicksman

"A masterful synthesis of modern medicine and engineering, revealing a human body brimming not only with biological information and ingenious molecular machines, but also with exquisitely engineered systems and subsystems that resemble but exceed the most advanced engineering techniques of our best engineers…. a most fantastic voyage of discovery."
—**Stephen C. Meyer**, PhD, *New York Times* bestselling author of *Darwin's Doubt*

"The authors are particularly good at raising the relevant questions about how all of this engineering came to be. Evidence of design and intent are everywhere…. I particularly enjoyed the way the authors tackled claims about 'botched design.'"
—**David Galloway**, MD; former President, Royal College of Physicians and Surgeons of Glasgow; author of *Design Dissected*

"A fascinating description… The human body is an elegantly engineered system, complete with intracellular nanotechnology, organ-level interconnectivity, and whole-body integration of countless delicate and precise systems."
—**Michael Egnor**, MD, Professor of Neurosurgery and Pediatrics at State University of New York, Stony Brook; named one of New York's best doctors by *New York Magazine*

"A brilliant tour of the mind-boggling interactive complexity of the human body. They make a compelling case that only an intelligent agent could possibly have accommodated the millions of design constraints needed to produce a living, breathing, moving, thinking, and reproducing organism like us."

—**William S. Harris**, PhD, Professor of Internal Medicine, Sanford School of Medicine, University of South Dakota; President of the Fatty Acid Research Institute

"A worthy and thorough addition to any person who seeks to discover, understand, and discuss the details of the human body's marvelous design."

—**Anthony Lyle Donaldson**, PhD, founding dean and Professor of Electrical and Computer Engineering, California Baptist University

"An excellent overview of the evidence demonstrating how the human body is the result of engineering design. The authors… also effectively refute arguments that there are many poorly designed elements of the human body."

—**Eric Cassell**, longtime engineering consultant for NASA and the FAA, author of *Animal Algorithms: Evolution and the Mysterious Origin of Ingenious Instincts*

"The authors show there is incomprehensible complexity upon complexity upon complexity within the human body... uncountable 'chicken-and-egg' scenarios found at all levels. Naturalism (Darwinism) fails miserably to explain these phenomena."

—**Geoffrey Simmons**, MD, author of *Billions of Missing Links*

CONTENTS

INTRODUCTION

I sing the body electric.
—WALT WHITMAN

THE HUMAN BODY IS AMAZING. EVEN A CURSORY LOOK SHOWS US that a lot is going on. Hands that wield a sledgehammer during the day can play evocative piano sonatas in the evening. In a triathlon, the same body swims, bicycles, and runs—three very different activities—in rapid succession and with extreme endurance. The same body that completed that triathlon can also climb a mountain (though perhaps on a different day).

Our bodies keep a constant internal temperature, manage our water levels effectively, and keep us going even when we eat the wrong foods.

When we stand up, our blood pressure adjusts almost instantly to keep blood flowing to the brain. We know when we need food and water. Even with our eyes closed, we can sense the position of all our body parts and make detailed adjustments in movement.

Our eyes distinguish the nuances across an amazing range of colors. The same eyes that work in painfully bright light can also see in almost total darkness. How do they turn light (photons) into information carried as electrical impulses, and how do our brains turn that into images?

Our ears face similar challenges, only they turn sound into electrical signals. Further, they're configured such that our minds can generate a three-dimensional understanding of the objects around us, just by the sounds those objects emit or block.

When we cut our finger, the blood quickly stops and the wound scabs over and heals. When we get sick, our bodies generally do an excellent job of fixing the problem and getting well again.

While our bodies are not the fastest, biggest, or strongest in the animal kingdom, they are the most versatile. The human body's range of capabilities boggles the mind.

On top of all this, we can make *new* people. Anyone who has experienced the birth of a child knows that in this astonishing process something special happens.

The human body is a marvel of engineering, but what is the source of the engineering? In 1859, Charles Darwin's *On the Origin of Species by Means of Natural Selection* purported to explain how new species, including the human species, arose without the benefit of a divine engineer. In the next century Darwin's theory was updated in what came to be known as neo-Darwinism. Now the random variations in Darwin's theory were understood to be random mutations in DNA. But could any series of unguided errors, over any period of time—with or without natural selection—achieve the wonders of the human body, or even the wonders of a single cell, itself wonderfully complex?

Not all causes are created equal. We can divide all causes into two main types: material causes and intelligent causes. Every kind of cause is either one or the other, or a combination of the two. In the end, then, the human body must have come about by some combination of forces from either or both these two kinds of causes.

Purely material causes involve the physical regularities of the universe, as described by the laws of mathematics and the laws and constants of physics and chemistry. Material causes are repeatable. The same inputs produce the same results. Their repeatability makes experimental science effective. But physical laws lack desire, intent, foresight, and planning, which limits their creative powers. No one would posit that a fully fueled, launch-ready Atlas rocket was generated by purely mindless material forces.

Purely material processes also encompass random events, such as random mutations to an organism's genetic information. And random events also lack desire, intent, foresight, and planning.

Unlike random mutations and other purely material processes, intelligent causes act with intention—they perform actions and build things to achieve intended goals and purposes. Intelligent agents visualize an outcome, plan how to achieve it, and execute the plan to make the vision reality. They make specific choices to achieve the desired outcome, guiding the construction, assembly, and activation of the end product. Intelligent agents generate information and give it meaning. They design systems that harness the laws of nature to perform tasks that nature could never otherwise do.

Natural causes must rely on blind trial and error, with lots and lots of luck. As we'll show, building any non-trivial system of systems is profoundly improbable, and achieving all the layers of coherent systems required for a functioning human body, in any time frame, is even more unlikely. We will make the case that the systems in the human body could only have been achieved through intelligent foresight and planning.

This book will benefit from two distinct, complementary perspectives:

- Physician Howard Glicksman's medical perspective— to help us look at some of the body's many intricate and interdependent systems.

- Steve Laufmann's engineering perspective—to help us appreciate the exquisite engineering of these systems: the mechanical, pneumatic, hydraulic, and electrical systems, the control systems, the internal signaling and coordination mechanisms, the information processing systems, and more.

Throughout, we'll base our observations and arguments on incontrovertible medical and engineering knowledge. The two of us have presented our case at much greater length, in a nearly 500-page book, *Your Designed Body*. What you hold in your hands is the much briefer version of that argument.

The view advanced in these pages does challenge the reigning paradigm for biological origins. But dominant paradigms aren't always the best paradigms. The history of science is replete with dominant paradigms that were overthrown when new evidence drove new theories to the fore.

In such cases, the champions of the dominant paradigms do not generally cede the field quickly or magnanimously. This is a central message of historian of science Thomas Kuhn's famous work *The Structure of Scientific Revolutions*. The Nobel Prize-winning physicist Max Planck put it this way: "A new scientific truth does not triumph by convincing its opponents and making them see the light, but rather because its opponents eventually die, and a new generation grows up that is familiar with it."[1] Or, as his point is often paraphrased informally, "Science advances one funeral at a time."

That's a bit more pessimistic than the reality. Already there have been some high-level public conversions to the design paradigm in the scientific and broader academic community, as well as a growing number of young scientists who are privately supportive but are keeping a low profile because they are at vulnerable points in their careers. Planck's observation, however, is true in the main.

Psychologist James Dobson tells a story from early in his career, when he worked in a clinic with patients who had varying levels of detachment from reality. One patient believed he had been dead for some time. Dobson tried everything he could think of to convince this poor guy that he was actually alive. Nothing worked. After much thought, he devised a foolproof approach. He asked, "Do dead men bleed?" The patient was outraged, "Of course dead men don't bleed. That's absurd." Dobson then pulled out a needle and pricked the man's finger. Staring at the drop of blood oozing from his skin, the man exclaimed, "Well, I'll be darned. Dead men *do* bleed."

The story illustrates a common foible of humans. When faced with evidence that challenges long-held assumptions, a person may resist letting go of the assumption that would be most reasonable to let go of, and instead will let go of another assumption, one he is less emotionally attached to.

As you examine the evidence laid out in these pages, our encouragement to you is, don't be the guy in the story. Be willing to follow the evidence wherever it leads.

1. Being Alive

Nature, in order to carry out the marvellous operations
in animals and plants, has been pleased to construct
their organized bodies with a very large number of
machines.... Machines will be eventually found not only
unknown to us but also unimaginable by our mind.
—Marcello Malpighi, Italian life scientist, 1697[1]

I (Steve) was still mostly a youth myself, yet there I was, waiting for my first daughter to come into the world. It was a long night of painful labor, but around seven in the morning, after much weeping and gnashing of teeth (mostly by me), out poked a little head and a new life entered the world.

It's wonderful to witness such a thing—a new human leaving the warmth and nurture of mother and beginning a new life of exploration and discovery. Only years later, after a couple more births, did I begin to explore just how special a new life is, and exactly how challenging life is from an engineering perspective.

The Greatest Discontinuity

We're surrounded by living things. Here on Earth, life flourishes almost everywhere we look, even in remarkably inhospitable places. Perhaps because life is so common in our world, it's easy to lose sight of how tenuous—in a sense, how unnatural—it is.

Some believe there must be a gradual continuum from non-living to "fully alive," but this is a presupposition starving for evidence. No one has ever seen something that was only partly alive. Alive but dying, yes. Partly alive, no. As biologist Michael Denton notes, "Between a living cell and the most highly ordered non-biological system... there is a chasm as vast and absolute as it is possible to conceive."[2] Or, as he likes to say in his live presentations, "Life is the greatest discontinuity in the universe."

In our experience, life always comes from life—never from non-life. No laboratory has observed life emerging from non-life, by any gradual or other process, and no one has any idea how to make something alive, even if all the parts are on hand. (Dr. Frankenstein apparently had it figured out, but sadly, his notebooks were lost in the fire.)

Life is set apart from non-life in many dramatic ways. For example, all living organisms, from the simplest bacteria to humans, must maintain an internal equilibrium distinct from the surrounding environment. This involves chemical makeup, physical organization, energy production and consumption, and many other properties. This is obviously a taller order for humans (with more interesting solutions) than it is for bacteria, but the essential problems are the same.

The fancy word for this property is *homeostasis*, which means "staying the same." Every living organism must maintain its separate equilibrium. If it can't, it dies. Equilibrium with the surrounding environment equals death.

For humans, homeostasis involves thousands of activities to maintain precise balances across dozens of chemicals (like water, oxygen, carbon dioxide, sugar, sodium, potassium, calcium, iron, copper, and manganese) across hundreds of processes to control energy, temperature, blood pressure, and many other life-critical factors.

The laws of physics and chemistry drive everything unrelentingly toward equilibrium with the environment. In contrast, life insists upon a separate and distinct equilibrium. This requires continuous energy and precise regulation in a complex and coherent choreography. Life must control its own outcomes in the face of forces working constantly to destroy it.

Life has another extraordinary capability. It can reproduce itself. Living things make copies of themselves that can make copies of themselves, and they do this from the inside (though sometimes with outside help, as from their mother). The copies must then become self-sufficient to the point that they, too, can reproduce. The number of physical problems that must be overcome to make this happen are far beyond what science currently understands, though in Chapters 2 and 3 we'll look at some impressive parts of the process we do understand.

No non-living object in the known universe can achieve both homeostasis and reproduction, and, notably, this includes anything designed and engineered by even the best human engineers.

The evidence tells us that these capabilities are required for life. They are *prerequisites* for life, not *outcomes* of it. There is no way for a creature to become alive first, then find a way to solve these problems. Without solutions to all these problems, life cannot exist.

At the Center of Life: The Cell

Your body is made of systems of systems of systems. At the foundational level are the trillions of cells that make up a human body. No living cells, no living humans.

In Darwin's day, a cell was thought to be a mere bag of chemicals with unknown function. Since then, science has shown that the cell is an extraordinarily complex factory—with its own information storage and processing facilities, energy production plants, and manufacturing plants for the thousands of structures and molecular-sized machines that perform the functions needed for life. Every cell must have all of the following:

Containment
Each human cell is enclosed by a thin, double-layer wall called the cell membrane. The membrane defines the boundaries of the cell, separating it from other cells and the outside world. It keeps what's needed inside and what's harmful outside. Obviously, the various chemicals and structures of the cell wouldn't be of much use if they could randomly wander off.

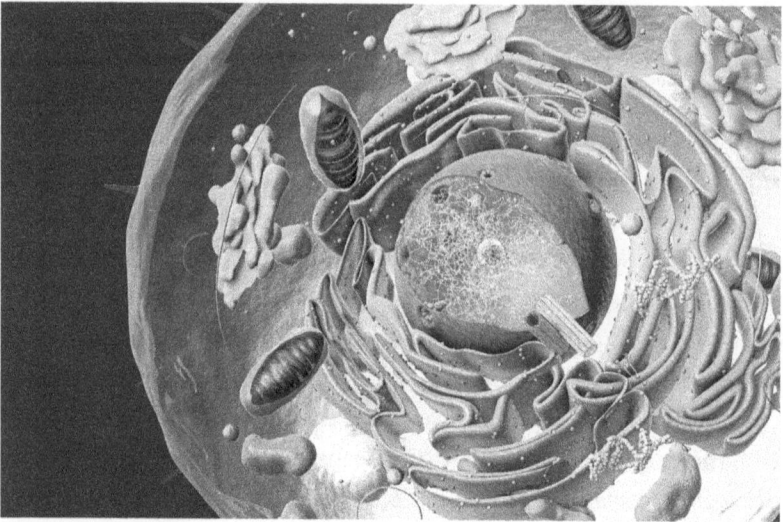

Figure 1.1. A typical human cell, showing cell membrane, cytoplasm, and organelles. As complex as the illustration appears, it is a vastly simplified representation. For a cell to be alive, it must solve numerous problems, and to do so it must contain an enormously complex, finely tuned factory.

Specialized Gates

For the cell membrane, a simple wall isn't enough. Just as a car needs gas and has to get rid of exhaust, each cell must bring in new supplies of the materials it needs, like oxygen, water, and sugar, and get rid of the toxic byproducts of its chemical reactions, like carbon dioxide and ammonia. So the cell membrane incorporates gates that are tuned to allow just the needed raw materials to enter the cell, and the toxic waste materials to be purged. These gates can be either active (like pumps) or passive (like pipes, typically with valves to control flow). The gates also must be specific enough to prevent the wrong materials from passing in either direction. In addition, some chemicals can easily pass through the cell membrane, but again, only chemicals needed for cellular viability, such as water.

The space within the cell membrane is filled with a fluid called cytoplasm, made up of water and dissolved chemicals like sodium and potassium. The volume of water inside the cell applies pressure against the cell membrane, just like the air in a balloon. The more water in the cell, the higher the volume and the more pressure it applies. And just as with too much air in a balloon, if there's too much water in the cell, the membrane will tear, killing the cell.

The water outside the cell, which surrounds and bathes it, contains many of the same chemicals, though most are at different concentrations from what's needed inside the cell. For example, the potassium level in the cytoplasm is much higher than in the water outside the cell, and the sodium level in the cytoplasm is much lower than in the water outside the cell. Maintaining this (and many other differences) at all times is critical for life.

Chemical Controls

Internal chemical balances must be maintained at all times, so the cell needs ways to control its internal levels of water, oxygen, carbon dioxide, glucose, hydrogen ions, sodium, and potassium, as well as other compounds for manufacturing amino acids, which are in turn necessary for DNA, RNA, and the various cellular machines. The right balances are critical and must be maintained within tight tolerances, regardless of the concentrations outside the cell.

Because concentrations inside and outside the cell are different, the cell must continually work against natural forces to maintain correct balances. This is a complicated problem. The cell has to control both its water content and chemical concentrations, but these affect each other. So the cell has to coordinate multiple controls to work together to maintain the right balances.

Structure

Merely having a cell membrane doesn't ensure a useable shape, so the cell also needs internal framing. Just as your body has a bony skeleton to support its organs, your cells have something called a cytoskeleton, made of intermediate filaments, microfilaments, and microtubules that together provide the structural parts inside the cell to support its shape. Interestingly, they double as "railroad tracks" for specialized transport proteins to travel along.

Modular Subunits

Compartmentalized subunits called organelles are suspended within the cytoplasm. The organelles house factories filled with molecular machinery to harvest and process raw materials, break down complex molecules into their parts, and build up other complex molecules from the raw materials.

A Transportation System

A human cell is large compared to its many millions of molecular parts, so it's necessary to transport materials from place to place within the cell. Materials simply floating from place to place would be too slow and inefficient.

Materials are loaded into sacs called vesicles, which are matched to motor proteins headed in the right direction. Motor proteins are specialized for about forty different types of cargo. Some motor proteins (kinesins) travel one way along the cytoskeleton, while others (dyneins) travel in the opposite direction, so the cell matches the right motor proteins for each payload and target destination.

In a meticulously choreographed operation, an organelle dumps outbound materials into a vesicle; then a motor protein picks it up and pulls it to its destination. There the target organelle membrane merges with the vesicle, in effect opening the cargo container and absorbing its contents.

How the cell knows what materials to carry where and when remains a mystery. But it's a good thing it all works, or you wouldn't be here to read about it.

Energy Production

Cells work hard, and that takes lots of energy. Energy-bearing chemicals must be converted to usable energy packets, which must then be consumed to perform the many tasks of maintaining life.

The mitochondria are your cell's energy factories. They use a series of specialized proteins to break down an energy-rich sugar, called glucose, in the presence of oxygen, to release and harness that energy. (And each molecular machine must have the right machinery to consume the available energy.)

Materials Production

If the cell had to rely on its environment for all the chemicals and complex molecules it needs, it wouldn't last long. Given the complex molecular machines that are essential to its life, the cell must have a way to convert the simpler chemicals it can get from its environment into the more complex chemical building blocks it needs to sustain life.

Information

The production of virtually everything in the cell is based on digital information, encoded in DNA and other complex molecules such as various RNAs.

The nucleus holds this information and is the "brains" of the cell. The nucleus of each of your cells contains twenty-three pairs of chromosomes, each made of long DNA molecules. Each DNA molecule consists of two separate strands of chemicals bonded together that spiral into a double helix shape. The DNA molecules contain much of the information for life—including specifications for the complex molecules and molecular machinery the cell needs.

Information Processing

Information is no good unless it can be accurately decoded and processed. And this takes a lot of specialized molecular machinery working in a precise choreography.

To get the information to make a protein or a duplicate copy of itself, the DNA molecule has to be "unzipped." Special enzymes called helicases break the hydrogen bonds that hold the double helix together, allowing the DNA molecule to unwind.

If the cell is making a protein, an enzyme called RNA polymerase jumps in to join nucleotides together to form something called messenger RNA (mRNA), which is a copy of the specification in the DNA. mRNA is transported from the nucleus (where the DNA is located), through the nuclear membrane, to a ribosome where other enzymes are used to assemble the protein.

In a separate process, when the cell needs to make an identical copy of the DNA molecule, it uses DNA polymerase and two other enzymes.

Proteins as the Cell's Workhorses

Much as cells are the core elements of the body, proteins are the core elements of the cell. Proteins are amazing, molecular-scale machines that construct, assemble, disassemble, join, cut, paste, copy, catalyze, enclose, and transport other molecules. Some proteins are essential for making energy. Others consume energy. Some do both. Proteins do almost all the work in your cells.

In the standard genetic code, proteins are made from an "alphabet" of twenty different amino acid "letters." Each letter is a complex molecule made of hydrogen, oxygen, carbon, and nitrogen (and in two cases, sulfur), and each has a unique physical configuration, with unique chemical and electrical properties. (A twenty-first amino acid, containing selenium and involved in different coding and translation mechanisms, is beyond the scope of this book.)

Proteins—along with peptides, which one can think of as miniproteins though with some differences—are made of a string of these amino acid "letters," linked together in a specified order unique to that particular protein or peptide. As in functional software code, not just any order of letters and symbols will do. Rather, a precise order is needed. Amino acids are connected end to end, like a chain, where the chain length varies from one peptide or protein type to the next, from three (for the glutathione peptide, the smallest known functional chain) to 34,350 (for titin, the longest known protein chain in the body). Human proteins average around 480 amino acids in length.

The precise order of amino acids in a protein is specified in the DNA that codes for that particular protein. DNA uses a four-character alphabet, with values represented by the letters T, C, A, and G. These letters are arranged to form three-letter words. These three-letter words are called codons and tell the translation machinery the type of amino acid to add. This set of rules is termed the genetic code. All but one of the amino acids are specified by at least two different codons, and there are also three-letter words/codons that mean "start" and "stop." As the DNA is processed linearly, the "start" and "stop" codons tell the translation machinery where to begin and end the linear chain of amino acids.

When the specific order of a string of amino acids is just right, the chemical properties and locations of the amino acids enable the protein to spontaneously fold into a stable three-dimensional shape. Sometimes proteins need help folding, provided by a special protein called a chaperone. The protein's three-dimensional shape, together with the relative locations of its various bumps, notches, crevasses, and electro-chemical attractions, determines the functions the particular protein can perform.

Experiments have revealed that functional protein shapes are extremely rare among the set of all possible amino acid sequences. The overwhelming majority of possible sequences will not fold into a stable protein shape and, therefore, are unlikely to provide a useful function. Multiple independent researchers have found this to be so.

For example, Douglas Axe, at labs in Cambridge, England, showed that for every DNA sequence that generates a relatively short (150 amino-acid) functional protein fold, there are about 10^{77} combinations of the same length that will not yield a stable, useful protein. That's one chance in the number 1 followed by 77 zeroes! To put those odds in perspective, imagine a blind man is sent on a cruel scavenger hunt: He must try to choose, at random, the one winning grain of sand from all the grains of sand in all the beaches and desert dunes on earth. Since there are roughly 10^{20} grains of sand for him to choose from, his odds of happening upon that one winning grain of sand are far, far better than one chance in 10^{77}. For all practical purposes, stumbling across a DNA sequence that generates a foldable protein of even moderate size is impossible.[3] And, of all the sequences that do fold into a stable shape, only a very few will perform a task that's useful to a given organism.

Illustrating this reality is the fact that while the potential medical benefits of designing new proteins are immensely attractive and therefore have garnered considerable funding and research efforts, progress in designing proteins that are both new and useful has nevertheless proven extraordinarily difficult. Inventing a useful new protein fold is a bit like finding a molecule-sized needle in a galaxy-sized haystack.

The obvious question is this: If it's so hard to intentionally design a single new functional protein, how could any non-intentional cause ever manage it, never mind doing it for the tens of thousands of different kinds of proteins in the human body (or in all the other kinds of organisms)?

One final illustration on the point: The situation is like trying to spell out a sentence with Scrabble letters. An intelligent agent can do it quickly, no problem. But try doing it by randomly dumping out a box of Scrabble letters and hoping to get a long, coherent English sentence just by accident. The odds are just too slim.

Enzyme Cascades

Most of the chemicals the body needs are fairly stable. They don't break down or interact with other chemicals well, or quickly. If you put them in a test tube and swirl them around, very little of interest will happen. This presents a problem for life, because almost all of life's essential processes require rapid changes in a variety of chemicals.

Some of the chemicals must be broken apart (catabolism). Others must be assembled (anabolism). Without help, such reactions would take days, years, or might never happen.

For example, a glucose molecule contains high-energy bonds, but these bonds are stable and not readily released under normal circumstances. The glucose molecule must be broken apart step by step to release its embedded energy. No cell could survive long without a rich supply of energy, so this reaction must happen quickly and predictably.

The energy in glucose is stored in stable molecular bonds so that it doesn't get randomly released into the bodily environment, or even into the gastrointestinal tract (to be assaulted by acids). Instead, it stays intact until exactly where and when it's needed—when it gets to the energy factory in an individual cell, which has exactly the needed machinery to release (and put to good use) the glucose molecule's energy.

Enzymes are a special class of proteins that come to the rescue. They greatly speed up (catalyze) certain reactions, which is what the cell needs for pretty much all its internal activities. Specific enzymes catalyze specific reactions, so the cell needs a lot of different enzyme types. For example, in human cells, processing glucose to release and use its energy involves three major parts: glycolysis, the citric acid (Krebs) cycle, and the electron transport chain. In total, it takes twenty-three distinct enzymes working cooperatively to process glucose. The molecule that results from the first catalyzed reaction becomes the input for the second, and so on to the twenty-third and final result.

An enzyme cascade is a bit like an assembly line where each worker has a specialized skill for his particular step in the assembly process. That worker and no other can perform that step. Switch the workers around and the assembly process grinds to a halt. Only when each step happens correctly, and in the right order, does the final product roll off the line.

Further complicating matters for glucose, each of the three major processes for extracting its energy happens in a different place in the cell, so transport is needed to move intermediate products between locations. Notably, some of the intermediate molecules are toxic, so the whole process must be tightly controlled to prevent toxic chemicals from building up and killing the cell.

Several of the best-known (and most toxic) human poisons work by interrupting a part of this central energy process. For example, cyanide blocks just a single enzyme in the chain, while arsenic blocks multiple enzymes in the chain. These poisons are especially nasty because every cell in the body uses this same process, so they will kill any cell they encounter. And when the cascade is interrupted at any point in the process, energy production fails and the cell dies. When enough cells die, the body dies.

The Cell's Protein Machinery

Human cells can make many different proteins. While it's believed that about 20,000 human genes code for proteins, in many cases a single gene can generate a potentially large number of different proteins, so no one really knows how many different proteins are in the body. Estimates range from a few tens of thousands to several million.[4]

But all these proteins are not needed in every cell at all times, so the cell uses internal controls to determine whether and when to build each of these proteins. Simply put, each cell determines when a particular protein is needed, and makes more of it, as needed. If this weren't the case, the cell would expend virtually all its energy and raw materials generating proteins that aren't needed.

The specific shape of a folded protein will expose some of its amino acids to nearby molecules. The specific mechanical and electrochemical properties of the protein provide a precise key-in-lock fit with one or more other specific molecules, such as molecular oxygen (O_2), glucose, an amino acid, or another complex protein shape such as a hormone or neurohormone. This points to thousands of chicken-or-egg problems evolutionists must confront. So, for example, which came first, the neurohormone or the neurohormone receptor?

Now let's look at one of the cell's many thousands of specialized protein machines—one of the best understood in the human body. This example should give at least a taste of how astonishing the human cell is when its thousands of different molecular machines are considered together.

ATP Synthase

As discussed above, human cells use twenty-three distinct enzymes to release the energy from a glucose molecule. The final, crucial enzyme in this chain is ATP synthase.

ATP synthase is an astoundingly complex, life-critical molecular machine that looks and works like a turbine. It's attached to the inner membrane of the mitochondrion and has all the parts of a human-designed rotary motor, including a rotor, stator, and camshaft. Protons flowing into a mitochondrion must pass through precisely shaped channels in the ATP synthase, causing it to spin. Below the motor are six large subunits. As the driveshaft spins, a bump on the camshaft causes these lower subunits to open and expose a specially shaped and charged crevice, allowing a molecule called adenosine diphosphate (ADP) to enter. As the shaft rotates, the protein causes an additional phosphate group (PO_4) to attach to the ADP, turning it into adenosine triphosphate (ATP), the primary energy molecule in the cell. As the shaft continues to rotate, the ATP molecule is ejected from the protein for use in one of the other molecular machines in the cell.

ATP synthase thus converts the natural flow of protons through this membrane into high energy phosphate bonds (as ATP) that the cell then uses to power its other protein machines.

A static illustration of ATP synthase can't do it justice. To grasp the exquisite design of this machine, watch animated videos of it in action, such as the short YouTube video "Molecular Machines—ATP Synthase: The Power Plant of the Cell."[5]

Each molecular machine that uses energy must incorporate a special ATPase domain, which extracts the energy from the third phosphate bond in ATP, converting it back into ADP, its lower-energy form.

Here's how it works: Each cell has about a million sodium/potassium pumps embedded in its cell membrane. These tiny molecular machines must work really hard, all the time, to keep the cell alive. Since the water inside the cell has a high concentration of potassium ions (K^+) and a low concentration of sodium ions (Na^+), and the water outside the cell has a low concentration of K^+ ions and a high concentration of Na^+ ions, diffusion makes these ions move across the cell membrane in the opposite directions—K^+ ions out of the cell, Na^+ ions into the cell—to try to equalize the concentrations on either side. In addition, the fluid inside the cell contains a high concentration of protein compared to the fluid outside the cell, so water must move into the cell by osmosis to equalize the concentrations of chemicals on either side of the membrane. Due to this rapid movement of water into the cell and long before the ion concentrations equalize, the cell will take in more water than its membrane can hold and will die by explosion. To prevent this, the cell uses a Na/K pump, which contains an ATPase domain (consuming the energy from ATP) to constantly force Na^+ ions out of the cell and bring K^+ ions back in—against their natural tendency to go in the opposite direction. This maintains the proper chemical balance and water volume.

When you're at rest, about one-quarter of your body's energy needs are taken up by these hard-working pumps. Since each of your cells has about a million of these pumps, and you have 30 trillion or more cells, all trying to maintain their chemical balance, that's

30,000,000,000,000,000,000 (3 x 10^{19}) sodium-potassium pumps working, even when you're sleeping.

One machine turns ADP into ATP; another turns ATP back into ADP—a complete energy cycle. One system infuses ADP with energy; another, encoded separately in the DNA, uses that energy. This cycle can ramp up production quickly, as needed, since nothing new needs to be manufactured—only rapidly recycled. Notice, too, that it takes two machines working together to achieve function. Neither would be useful without the other.

If that's not challenge enough for any causal theory of origins, here's another: Some of the enzymes needed to extract ATP from glucose must consume energy (in the form of ATP) to perform the chemical change at their step in the process. Thus, to create ATP as output, ATP is required as input. There is no other way known to make ATP except by consuming ATP, and this presents a quandary known as causal circularity. The product of the reaction is required to start the reaction itself. So where do you get the ATP required to make the first ATP? There are many similar examples in biology.

Engineering Hurdles

To be alive, each cell must perform thousands of complicated tasks. This includes containment, the operation of special-purpose gates, chemical sensing and controls (for many different chemicals), supply chain and transport, energy production and use, materials production, and information processing.

What's required to make all this work? Designing solutions to problems like this is hard, especially given two additional requirements.

The first, *orchestration*, means the cell has to get all the right things done in the right order at the right times. The activities of millions of parts must be coordinated. To this end, the cell actively sequences activities, signals various parts about what to do, starts and stops various machinery, and monitors progress.

The second requirement is *reproduction*. As if being alive weren't difficult enough, some of the body's cells need to generate new cells. This imposes a daunting set of additional design problems. Each new cell needs a high-fidelity copy of the parent cell's internal information, all the molecular machines needed for life, and a copy of the cell's structure, including the organelles and microtubules. And it needs to know which internal operating system it should use. Once these are all in place, the cell walls must constrict to complete the enclosure for the new cell, without allowing the internals to spill out.

Somehow cells solve all these problems. Each cell is a vast system of systems, with millions of components, machines, and processes, all of it necessarily coherent, interdependent, tightly coordinated, and precision-tuned if the cell is to be alive rather than dead.

To better grasp the engineering sophistication of human cells, we encourage you to watch a four-minute online video by the Human Protein Atlas titled "The Human Cell."[6] (At some point this video will be superseded by a new and improved video on the subject, so interested readers are also encouraged to search the web for newer resources.)

When Darwin first espoused his theory, none of the cell's complexity was known, so the idea of making a cell seemed simple. If the cell merely contains "jelly-like protoplasm," well, how hard could it be for unguided natural processes to make that? Science and medicine (and engineering) have exploded that simplistic view of the cell.

In fact, we now know that being alive and reproducing, even just at the cellular level, are the two hardest engineering problems in the universe. Yet we're constantly told that these things originated without intention, through a long series of accidents. But there remains no plausible, causally adequate hypothesis for how any series of accidents—no matter how lucky and no matter how much time is given—could accomplish such things.

Presently it even lies beyond the reach of our brightest human designers to create anything of this kind. Human engineers have no idea how to match the scope, precision, and efficiencies of even a single cell, much less organisms composed of many cellular systems of systems, each system composed of millions or billions of cells.

2. FROM 1 CELL TO TRILLIONS

The complexity of these, the mathematical models, and how these things are indeed done are beyond human comprehension. Even though I'm a mathematician, I look at this with a marvel of, how do these instruction sets not make these mistakes as they build what is us?
—ALEXANDER TSIARAS[1]

IN A LONG CAREER, I (STEVE) HAVE WORKED AS A RESEARCH SCIENTIST, software developer, manager, and systems architect; in everything from research labs to telecommunications to government to manufacturing to hospital groups; with Fortune 50 companies with tens of thousands of employees and start-ups with just a couple of people. I've worked on systems of systems composed of thousands of systems that were independently developed, using different technologies from different eras, systems that address different goals based on different approaches and with vastly different architectures.

But I never once encountered a system that managed its own fabrication, assembly, and operations, from start to finish, until I turned my attention to living systems. Somehow, living systems know how to do these things.

For the human body, the process of development begins with a single fertilized cell and ends with a fully functional adult—growing from one cell to around two trillion at birth, then to around thirty trillion at maturity. But it's not the number of cells that makes this so impressive. It's the number of problems the developing body needs to solve along the way.

After conception the number of cells in the embryo begins to multiply and, before long, to differentiate into the hundreds of different cell types the body will need. About two weeks after conception, the embryo separates into three distinct germinal layers, shown in Figure 2.1: the endoderm (inside layer), mesoderm (middle layer), and ectoderm (outer layer). Eventually the cells in each layer differentiate into the body's many cell types, and later into the specific tissues and organs that compose the body's eleven major systems.

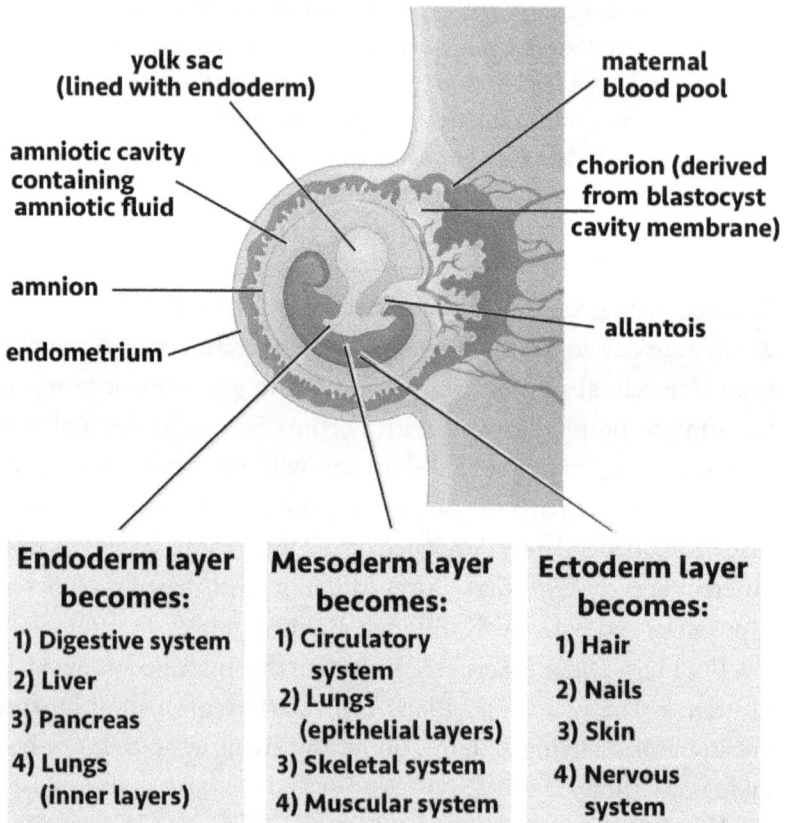

Endoderm layer becomes:	Mesoderm layer becomes:	Ectoderm layer becomes:
1) Digestive system	1) Circulatory system	1) Hair
2) Liver	2) Lungs (epithelial layers)	2) Nails
3) Pancreas	3) Skeletal system	3) Skin
4) Lungs (inner layers)	4) Muscular system	4) Nervous system

Figure 2.1. The three germinal layers of a developing embryo.

For this differentiation process to work, order and timing are everything. Much of what the baby will need at birth will require the full nine months of gestation to build. Some body systems can wait till near the end of the gestation process. Others must arrive much sooner. Which layers would you guess need to develop first if the embryo is to survive?

The cells of the endoderm will form the lungs and digestive system. But those aren't needed right away, since the mother has the duties of the lungs and digestive system covered via the placenta.

The cells of the ectoderm differentiate to form the skin and nervous system. Once outside the womb, they will allow this new human life to interact with and be protected from his surroundings. But for now, he's safe in the mother's womb, so he can do without this as well for now.

Some of the cells of the mesoderm will differentiate to form the musculoskeletal system. Once outside the womb, this system will allow this new person to move around and handle things. But for now, where does he need to go and what does he need to do? His job is just to sit tight and let his mother provide him with what he needs to grow and develop.

Other cells of the mesoderm will differentiate to form the kidneys. Once outside the womb, they'll allow the baby to control his salt and water content. But for now, his mother, through the placenta, has that covered too.

So what's left?

The embryo quickly transitions to using the placenta for nutrition. But getting nutrients is not enough. Those nutrients must be distributed to every cell in this new little body as it grows. For this the embryo needs a cardiovascular system—a heart, blood vessels, and blood—and these also come from the mesoderm.

At about three weeks, the new heart begins pumping blood through its developing circulatory system. Over the next several weeks and months, as the embryo becomes a fetus and grows and develops, the heart provides the driving force for life.

Forty Weeks to a Breathing Body

The process of building a new body from a single cell is extraordinarily complex. The timeline is short and the stakes are do-or-die. Below are some of the highlights of the process, and when they occur. T stands for the moment of conception.

- **T**: A one-celled zygote is created—a unique human with unique DNA.

- **T + 30 hours**: The zygote divides into two identical cells.

- **T + 3 days**: With further cell multiplication, this new life is now a sixteen-celled morula. The morula drifts through the fallopian tube on the way to the uterus.

- **T + 5 days**: With further dividing, the new life is now a blastocyst, and enters the uterus.

- **T + 8 days**: The blastocyst attaches to and implants into the lush endometrial lining of the uterus.

- **T + 2 weeks**: The blastocyst is now an embryo with an amniotic fluid cushion, yolk sac for food, and different cell types.

- **T + 4 weeks**: The heart has been pumping blood through the tiny body for several days, providing nutrients to immature organs and the early skeletal system.

- **T + 6 weeks**: The embryo has brain waves, hands with fingers, and feet with toes, and the airways are forming.

- **T + 8 weeks**: As the embryo enters the fetal stage, the yolk sac gives way to the placenta as the umbilical cord develops.

- **T + 10 weeks**: The fetus begins moving his fingers and toes, the intestines are developing, and the skin is covered with fuzz.

- **T + 12 weeks**: The kidneys are functioning, sending urine into the amniotic fluid. (The first trimester is completed at thirteen weeks.)

- **T + 14 weeks**: A week into the second trimester, the fetus makes breathing movements and begins sucking his thumb.

- **T + 16 weeks**: Now the fetus is growing more quickly. Bones harden, lung sacs appear, and the nerves start getting their myelin insulation.

- **T + 18 weeks**: The brain begins developing, differentiating into distinct areas for senses like vision, hearing, smell, taste, and touch.

- **T + 20 weeks**: The fetus has hair and fingerprints, and starts kicking as the heart and blood vessels fully develop.

- **T + 22 weeks**: Except for the lungs, the organs have developed well enough for quasi-independent living.

- **T + 24 weeks**: The bone marrow is making red blood cells. (Before this they were produced in the yolk sac, liver, and spleen.) The fetus is getting fatter and his intestines are moving.

- **T + 28 weeks**: A week into the third trimester, the baby is opening and closing his eyes. He sleeps and wakes regularly.

- **T + 32 weeks**: The baby puts on more fat and weight, his skull grows to fit his brain, and his lungs continue to develop.

- **T + 36 weeks**: Growth continues with more fat, bone hardening, and more development of the brain and lungs.

- **T + 40 weeks**: All the organ systems are ready, so it's time for this little one to head out, take his first breath, and meet his mother and father.

Birth and Transition

What triggers labor is not fully understood. Apparently, the mother's body detects and reacts to what's happening in her womb. At the same time, the baby sends signals that tell the mother's body that he's ready to be born.

Progesterone, the pregnancy hormone, maintains the lining of the uterus for the placenta. Also, throughout the pregnancy it keeps the uterine muscle relaxed so it doesn't contract and push the baby out prematurely. The continued growth of the baby begins to stretch the uterine muscle, making it more excitable.

As the time gets close, the aging placenta struggles to keep up with the growing demands of the baby. This stresses the baby, so the placenta sends out more cortisol, a hormone critical for lung develop-ment. Cortisol also converts progesterone into estrogen, which makes the uterine muscle even more excitable.

The combined effects of maximum uterine stretching, placental fatigue, the associated rise in cortisol, and the rise in estrogen and reduction in progesterone seem to cause the uterine muscle to contract and bring on labor. As this happens, the mother's body releases other chemicals, like prostaglandins and oxytocin, which strengthen the contractions and soften the cervix for the delivery of the baby.

Note the many ways that coordination occurs between the mother and her developing child. Their two separate bodies communicate with each other to orchestrate their activities. The programming that makes this system work in the mother was present when she herself was in the womb, when her body as an embryo was interacting with her mother's body. Several chicken-or-egg sorts of questions come to mind.

Once the baby is born, his body must take over all the functions he relied on his mother for, including respiration. This transition requires activating the lungs and rerouting the blood, as well as many other tasks like managing water and salt concentrations, fluid balance, and temperature regulation. The newborn's immune system must also be ready to go as the child enters this new and dangerous world.

And all this needs to happen quickly.

A Plan for Unfolding the Body Plan

The human body somehow finds a way to grow from a single-celled zygote to a thirty-trillion-celled mature adult. To accomplish a task of this magnitude requires a plan, including an order of engagement, together with many specialized systems whose jobs are to build out the body (as opposed to operating it once it's built).

The plan, together with its programming and build-specific mechanisms, must address each of the millions of problems above. Anything less would result in a body that doesn't function well or, more likely, doesn't function at all.

Given how precisely the body must do these things to sustain life, it's quite predictable, from an engineering perspective, that the process is extremely sensitive to errors. Even a small error in the development process likely will lead to a major problem in the outcome. This is generally true for systems of this kind.

The body is arranged in a hierarchy of layers—a design hierarchy from whole body systems down to organ systems to tissues to cells, and all the way down to organelles within the cells and their individual molecular machines. The development process must build out these layers as it goes, differentiating, assembling, and growing. Hundreds of facets of the design must be realized, from plan to body, through this process. We group these into four categories of build plans:

Differentiation

Well over two hundred specialized cell types must be made (and possibly several hundred more that have been theorized but are yet to be discovered and cataloged). All contain the same DNA as the zygote (excepting red blood cells, which lose their nuclei along with their DNA)[2], but each expresses a different program and performs different functions.

This means that as the new life takes shape, some cells will divide but won't make new copies of themselves; they will make something different—a cell with a particular program selected, which will turn it into a specialized cell. This process may involve various facets of the cell, including its shape, capabilities, and role(s) in the body.

Different cell types express different genes at different times, and in different organelles within the cell, to build exactly the proteins and enzymes needed for their specific functions. Only those genes needed for their specific function are turned on. This is remarkably efficient. Different cell types run different internal programs, or operating systems. They respond differently to different signals and stimuli, and they perform different tasks for the body.

But how does cell differentiation cause certain cells to run different programs? How does the body know which cells need which programs? Where do these programs come from, and where (and how) are they encoded? Remember that the body must control tens of thousands of different proteins, in hundreds of specialized cell types. How could programming of this complexity have come to be?

The body knows what machinery is needed to perform what jobs at what times in what areas of what cells. It's brilliant at producing only what it needs, and only when and where needed. The level of control required to pull this off is mind-boggling.

Further, these cells make up the various tissues and body parts, like organs and bones, and these are also highly specialized. Specialization of body parts includes shape, size, and materials. The teeth are made of very different substances than the eyes or the spleen. In fact, specialization of parts extends through the layers of the hierarchy, from the encoding of information to the body as a whole.

Organization

The specialized cells are organized according to the body's design hierarchy into tissues, organs, and body systems. The placement of various cell types in the body is precisely controlled. In our book *Your Designed Body*, we show many examples of this, from the dispersion of light-sensitive cells in the retina, to the distribution of different types of cells in the skin, to the specific positions of the bone cells that build and maintain the precise shapes of bones. As a lesser-known example, most of the tissues in your body incorporate various types of immune cells, distributed among the cells that perform that tissue's functions.

Millions of body parts are made by these cells, together with the extracellular materials that hold them together (e.g., the extracellular matrix, bones, and cartilage). Each must be precisely correct or bad things happen. Further, each part can perform its function properly only when it's in the right position with respect to other parts of the body.

Integration

The parts of the body must be integrated with precision. Each part must be held in place. Each must have the right connections and interfaces. These may be mechanical, electrical, fluid, chemical, or any combination.

Coordination

If the body is to work properly, each of its systems and parts must be properly tuned to the controls, timers, and signals it will need during its lifetime, together with the logic needed to achieve proper function.

The Parts of the Plan

Each of these four plans is quite challenging by itself. But to achieve them all is extraordinarily difficult. They must be properly "unfolded" and interwoven during development, starting from the zygote's one cell. The scope of this problem is daunting.

Clearly, all this must result from the body's higher-level development plan, so what does this plan contain? What information and programming are involved? Below we will consider what all must be there.

There are two parts to any plan of this kind: (1) the information that specifies the outcome (the *specifications*), and (2) the orchestrations, choreography, and processes for building out the plan (the assembly *instructions*).

The body, recall, is made up of individual cells (along with extracellular material, like collagen). Each cell is an autonomous entity that controls its own resources, makes and consumes its own energy, and has its own goals and the programming to achieve them. At each step in the development process, these cells are doing the work. How do they know where they are? How do they know how far along in the development process they are? How do they determine what to do next?

There are two possible approaches: *orchestration*, wherein the cells get instructions from a controller external to the cell; and *choreography*, wherein each cell determines its own course of action, based on its internal perception of the situation around it, combined with its internal programming.

If it's orchestration, where do the instructions come from? What systems track the development process and decide when and where to send out these instructions? If it's choreography, how do cells perceive the situation around them? What possibilities and differences can they distinguish, and how do they turn these into actions?

Most complex systems use a combination of these approaches. How the body does this is not yet known. But we can apply engineering knowledge to understand what must be there for the specifications and instructions to work, even if we have little idea where these things might be, or how they're encoded.

Specifications

DNA contains the exact specifications for the tens of thousands of different proteins the body must produce. But what else must be specified?

Clearly, the target state also must be specified. This includes what the parts are, where they belong, and how they're interrelated. It includes all the specializations, organizations, integrations, and coordination discussed above.

But much more is required. Coordinate systems are needed to manage and inform the cells about their placement and their relative positions, such that they're able to form incredibly complex 3-D shapes, like those of the vertebrae or the bones of the middle ear (ossicles), each of which must have a distinct and precise shape.

The body needs axes for lateral (left to right), longitudinal (head to toe), and depth (front to back). Placement of the parts along these axes must be precise. There is structural symmetry on the lateral axis, so we need systems to maintain size and length consistency on either side. (It's helpful when both legs are the same length, and when the knee joints are in the same position along the longitudinal axis.) At the same time, many of the body's internal organs do not have lateral symmetry. The heart and the spleen are always on the left side, and the liver on the right. This also must be precisely specified.

Also included are specifications for hundreds, or more likely thousands, of control systems, through all layers of the design hierarchy. Each control system must have at least three key parts—sensors to detect what needs to be controlled, comparators to apply decision logic, and actuators to change the thing being controlled. Without all three, control would fail and life would cease. Each requires finely tuned setpoints and thresholds. Many of these must also be positioned in just the right places. How does a growing fetus know to put the special CO_2 sensors in the brainstem, and the O_2 sensors in the aortic arch?

Build Instructions

The body also needs a set of do-it-yourself instructions, an action plan for building the body. These tell the developing body what, when, where, and how to do it. These instructions tell the cells when to divide and differentiate. They instruct the dispersion of different cell types in different tissues and organs. And they define the routing and placement of the piece-parts, like blood vessels and nerve bundles. Success requires at least all of the following:

- *Fabrication instructions*, which define how to build individual parts and what materials to use.

- *Assembly instructions*, which define the precise order in which the parts and systems will be built.

- *Launch instructions*, which define the stages at which various body systems, like the heart, will start performing their function.

- *A schedule*, which understands (monitors?) the body's progress and keeps the whole works on track—in effect, keeping the project on time and within budget.

Testing and Course Corrections

Some form of feedback is required during development. For human bodies to be made with so few errors, there are likely many subsystems to detect the current status (and perhaps also the trajectory) of the development process, compare it to the target plan, and make course corrections.

Much about how the body manages all of the above remains a mystery. But engineers can predict that such things must exist and will one day be discovered.

Making the Whole

Each of the body systems and mechanisms discussed in this book presents hard problems to that tiny zygote. If it's ever to become a mature adult, it needs to solve a lot of tough problems. There is nothing random about how these things must work if a mature human is to result.

Unfortunately for the curious, most of the specifications and instructions above are written in languages the scientific community does not yet really understand, and implemented in microscopic, molecular details that are hard to decipher. These appear to be "compiled" into thousands of gene regulatory networks (GRNs) and possibly also various epigenetic systems, but at that level of detail they are extremely tough to unravel. It's a bit like trying to figure out what a computer program is doing by watching the values change in the computer CPU's registers. The observational data are at too low a level to make much sense. Sadly, we don't have access to the source code, so we're left with the long and tedious process of deciphering the signals at ground level and reverse engineering to determine the underlying logic.

So, much of how development works remains a mystery. But we're thankful that it does work, since it means we get to exist.

A Few Thoughts

Millions of new people are born every day. Because it's so common, it's tempting to conclude that it must also be easy. But nothing could be further from the truth. The making of a new human is perhaps the most impressive sequence of events in the observable universe. It's the closest thing we have to a repeatable miracle.

When something goes wrong in development, we're prone to ask why, especially when it happens to the child of a friend or family member. It's a reasonable question, but we run the risk of passing over the more profound question: Why does anything in the process ever go right? After all:

- The single-celled zygote has nine months to build a complete, "standalone" human body composed of roughly two trillion cells.

- It must differentiate into hundreds of specialized cell types. Each differentiation must happen in the right place at the right time.

- Each specialized cell must be placed in just the right location to achieve its special function(s).

- Each part must be made of the right materials and have the right shape.

- Cells must be organized into tissues and organs, according to the body's design hierarchy. Different tissues have different types of cells, dispersed in specified positions. Tissues and organs must be organized into body systems. Body systems must be organized into a whole body.

- Thousands of control systems must be generated.

- Each body system must be given the right capacities, parameters, and setpoints to operate effectively.

- All the above must be done in the right times and places, such that the end result is complete and coherent.

- The process of fabricating and assembling parts must interleave systems to achieve the needed interdependencies.

And all this must be done from the inside. There is no external factory, and no external tools are involved—though, of course, the mother's body is providing raw materials and removing waste materials.

Building a self-sufficient human body from a single-celled zygote requires solutions to millions (billions? trillions?) of problems. The sheer complexity of this challenge, coupled with the beauty of the orchestrations involved, is breathtaking.

3. Fuel for Baby

*The two most important days in your life are the day you are born
and the day you find out why.*
—Anonymous

ALL CELLS NEED OXYGEN AND NUTRIENTS. EARLY LIFE IS NO
exception. Fertilization results in a single-celled zygote, which
multiplies through cell division to become an embryo. In the early
phase, the embryo gets what it needs by diffusion from the fluid
around it. This works when there are only a few dozen cells. But
within several weeks the embryo will grow into a fetus, and in a few
months into a newborn with around two trillion specialized cells,
organized into coherent, interdependent, finely tuned organ systems.

For this to be possible, the embryo needs a good way to get oxygen and nutrients, and to get rid of carbon dioxide and waste materials. If the baby cannot meet this challenge, he will not survive.
But he's in a special situation, dwelling inside his mother, so he'll
need a solution altogether different from anything else in the body's
inventory—a distinct yet temporary system that can meet this need
while he's developing his permanent internal systems. The baby's
growing body must generate these systems coherently, along with
their interdependencies, essentially from scratch, and remain alive at
all times while doing so. How do you build a series of finely tuned,
coherent, interdependent systems, each necessary for life, and stay
alive the whole time? It just wouldn't do if the body needed to go dead
for a while, build some stuff, then come back to life when everything
was ready to go.

This is an extraordinarily difficult challenge. No human engineer has ever designed a system capable of anything remotely like this. To say the body's solutions are brilliant is a bit of an understatement. This is genius engineering.

Early Life's Essential Supply Chain

What the child in the womb needs is a complete set of temporary systems to meet the respiratory needs of his rapidly growing body, to keep him alive until his own systems are ready to take over. This means that, in parallel with the developing child, other distinct systems must be developed, put into use, and grown as the developing child's needs increase. Then at birth, when they are no longer needed, these systems must be discarded as the child transitions to long-term systems.

These systems are functionally analogous to scaffolding in a building project—a temporary structure put in place to provide support while the building's permanent structural elements are assembled. Of course, the scaffolding for life in the womb is a significantly greater engineering challenge, involving a fresh set of interwoven hard problems, which require solutions that will take yet another distinct set of mechanisms and orchestrations.

The first step in this scaffolding starts when the growing embryo contacts the lining of the uterus. The embryo's outer tissue (trophoblastic tissue) emits enzymes that cause cells in the lining of the uterus to die. This process is not well understood, but it appears to work in two different ways. The embryo's enzymes break down the walls of some of the cells in the uterine lining (from the outside, called lysis), and instruct some of the cells to self-destruct from the inside (apoptosis). Both are fascinating processes. How do the enzymes target only the cells in the uterine lining, and not the embryo's own cells? No one really knows, but it's a good thing it works.

As the lining walls break down, the embryo gains an entry point into the nutrient-rich uterine lining—a place to implant. It attaches and burrows in further. For the next few weeks the embryo gets its oxygen and nutrients from the cells lining the uterus.

The embryo's DNA differs markedly from the mother's, and yet in the great majority of cases the mother's body recognizes that this microscopic new person is not a foreign attacker, so her body doesn't deploy its defense mechanisms to kill the baby. No one really knows how this works or what mechanisms are involved, but the mother's tolerance for her baby's body is recognized in the medical community as one of the great wonders of pregnancy, without which none of us would be here.

Placenta to the Rescue

As the embryo develops further, he continues making enzymes and breaking down more of the uterine lining. Eventually the embryo's outer tissue joins with the lining of the uterus to form the placenta. The placenta, shown in Figure 3.1, is a specialized organ that brings the developing child's blood into close proximity with the mother's blood, permitting the chemical building blocks of life to move back and forth between the mother's and the baby's circulation while keeping their blood separate. The close-up on the left side of Figure 3.1 shows how this works. The mother's body supplies the oxygen and nutrients her child needs and gets rid of the waste products that would kill him.

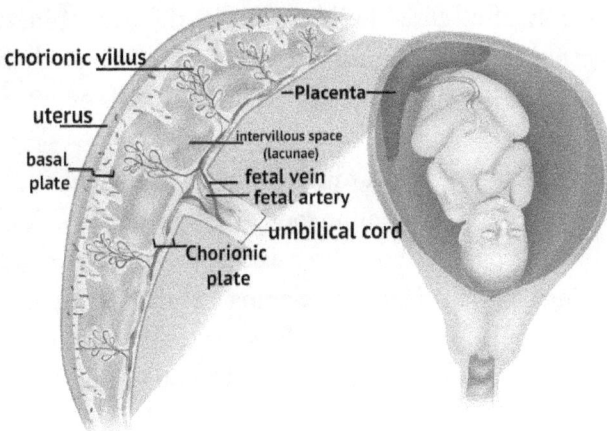

Figure 3.1. The microscopic view (left) of the fetal and maternal interface within the placenta. The placenta (right, above the baby) grows into an organ separate from the fetus or the mother.

This may be the most visually obvious example of the coordination required between the mother and her developing child, in which their two separate and genetically distinct bodies coordinate their activities in precise ways. In this case, tissues of the embryo combine with tissues of the mother to make the placenta—a totally separate organ that provides the scaffolding needed to keep the developing child alive. The placenta enables the mother to sustain the developing child while his internal organ systems and tissues are being fabricated, integrated, and launched. The developing child is, quite literally, on life support between the zygote phase and birth, when his body is finally ready to take over the job.

The placenta has no purpose aside from the development of the new life in the mother's womb, so it separates from the mother's uterus and is expelled from her body right after the baby's birth. The placenta doesn't get the credit it deserves, probably because it pales in comparison to the impressiveness (and excitement) of a newborn baby. But it invites a question: How does the placenta, a complex two-person system, come to be when that system is required for either individual to exist?

Without the placenta, it would be game over for the human race. Moreover, the placenta is just the first step in the continuous and increasing supply of essential building blocks the developing child needs. Oxygen and nutrients must be distributed to every cell in this little body, which grows by continually adding new cells over its nine-month gestation.

For this the baby needs a cardiovascular system—a heart, blood vessels, and blood—to come online soon after fertilization. And this is exactly what happens. At about three weeks, the embryo's new heart begins pumping blood through a developing circulatory system. Over the next eight months, as the embryo becomes a fetus and continues to grow and build out systems, his heart and blood vessels must grow with him to keep all his other cells alive and healthy.

As the preborn baby adds more and more cells, soon it's impossible for each cell to get what it needs by diffusion from the external environment. A more effective system is needed. Happily, the heart develops enough to start its work, circulating nutrients throughout the body, very early in the lifecycle, just in time to meet this rising demand.

But this is months before the brain's autonomic control system is ready to manage the heart's activities. How does the heart do its job without the brain in place to control it? The heart has its own internal pacemaker, so it will keep ticking even in the absence of the brain's autonomic control system. This is sufficient early in development, because during this phase the heart doesn't need to respond to rapidly changing metabolic needs in the child's body. After birth, by contrast, the brain's autonomic control system is indispensable, where it compensates for changes in activity levels, ambient temperature, fever due to infection, and other demands the body will place on the heart.

It's impressive that each part of the body is built exactly when it's needed, and in exactly the order needed.

A Survival Dilemma

The developing embryo's need for a cardiovascular system presents yet another problem that must be solved if he is to survive for more than a few weeks. As we will see in Chapter 4, the blood cycles through two loops in a sort of figure-8 pattern. The right side of the heart pumps blood through the pulmonary artery to the lungs where it picks up a new supply of oxygen. The oxygenated blood then returns through the pulmonary veins to the left side of the heart. From here the blood is pumped through the aorta and systemic arteries to deliver oxygen to the tissues throughout the body. Once the blood has given up some of its oxygen to the tissues, it cycles back through the systemic veins to the right side of the heart. This cycle repeats as long as you live. When it stops, you stop.

Heart Anatomy

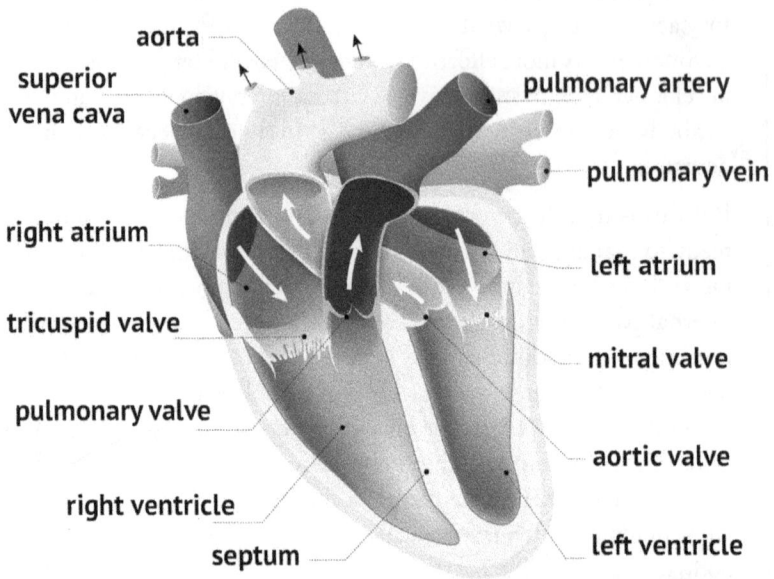

Figure 3.2. The chambers and valves of the human heart.

As with normal circulation, the fetus cycles blood from the left side of the heart through systemic arteries to the tissues. However, because a fetus's lungs don't have access to the outside air, oxygen must come from somewhere else—from the mother, via the placenta.

Blood pumped from the fetal left ventricle travels in two umbilical arteries to the placenta, where it picks up oxygen derived from the mother's lungs. The oxygenated blood then returns from the placenta to the right side of the fetal heart through its umbilical vein.

The fetus must get this oxygenated blood to the tissues. But the pathway needed after birth—through the lungs, back to the left side of the heart, and then on to the tissues (the pulmonary arterial system)—is not yet open for business. While the placenta has provided the necessary oxygen, the fetus can't send oxygenated blood through systemic arteries to the tissues they supply.

So the fetus needs a different circulatory pattern while in the womb. And with an alternative solution in place, lungs can delay developing until closer to birth. In fact, the baby's lungs won't make enough surfactant in the alveoli to allow that first breath until just before birth (see below for more on surfactant). This is why doctors have to work so hard with premature babies to establish and maintain their breathing.

A Survival Solution

The heart's circulatory pathway through the lungs is essentially closed for business during fetal development, so it cannot send blood from its right side to its left side via the lungs, as it will have to do after birth. Instead, the body employs a clever solution: two shunts (openings) that divert blood from the right side of the heart so as to bypass the baby's "not-yet-ready-for-prime-time" lungs and enter the systemic arteries to supply the body's tissues with oxygen.

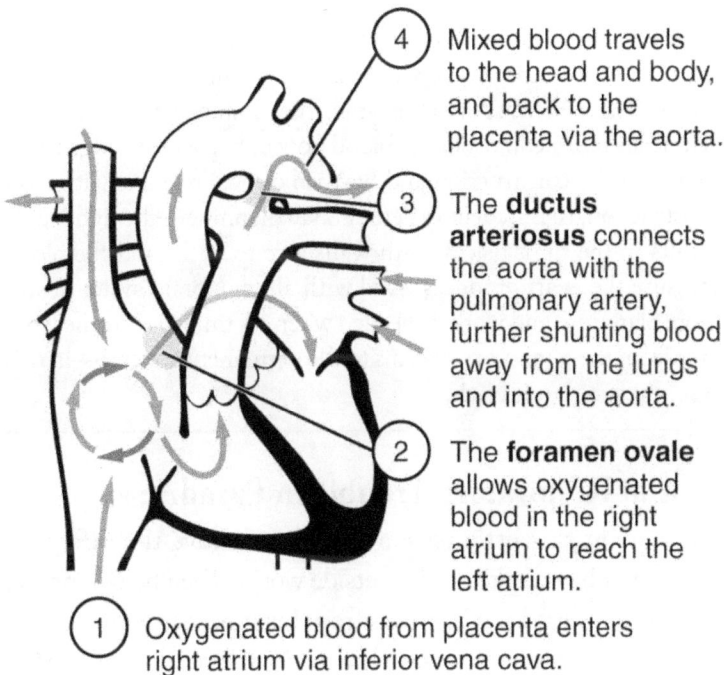

(4) Mixed blood travels to the head and body, and back to the placenta via the aorta.

(3) The **ductus arteriosus** connects the aorta with the pulmonary artery, further shunting blood away from the lungs and into the aorta.

(2) The **foramen ovale** allows oxygenated blood in the right atrium to reach the left atrium.

(1) Oxygenated blood from placenta enters right atrium via inferior vena cava.

Figure 3.3. The blood's path through the heart during fetal development.

The first shunt is called the foramen ovale. (See Figure 3.3.) This is a flap-like opening in the wall between the right and left atria. Since the lungs are closed and no blood is going from the pulmonary veins into the left atrium, the blood pressure here is lower than in the right atrium. Hence, this flap acts as a passive valve, diverting blood from the right atrium to the left, from which it's pumped into the left ventricle, then through the aorta and the systemic arteries to the rest of the body.

The second shunt is called the ductus arteriosus. (See Figure 3.3.) This is a pipe-like passageway between the pulmonary artery and the aorta. Because the lungs are closed, there is high resistance to blood flow in the pulmonary artery, so its blood pressure is higher than that in the aorta. This naturally diverts blood from the pulmonary artery to the aorta and from there to the rest of the body through the systemic arteries.

When a road is under construction but traffic still needs to get through, it's common to deploy an alternate but temporary route, a detour. The fetal heart does the same. The foramen ovale and ductus arteriosus redirect most of the blood flow to bypass the lungs while they're under construction and have no critical role. As with road construction, the circulatory detour takes planning—the right-sized shunts in the right places, with walls able to withstand the blood pressure the heart produces. And with all temporary systems, still more planning (and work) is needed when it's time to shift the work over from the temporary systems to the permanent systems—in this case, the lungs—at birth.

Potential Respiratory Trouble in Paradise

At some point this little person is going to leave the safety of his mother's womb and take on the outside world. Then he will no longer be able to get everything he needs from his mother's body. So among the baby's many other birthday chores, he will need to breathe. This fact raises yet another challenge for the developing child.

The transition from inside the womb to outside is critical, as the child can only live a few minutes without oxygen. Those previously useless lungs need to come online and work, and they need to do so more or less immediately.

When you fall into an ice-cold alpine lake, your body's first instinct is to inhale deeply. That instinct may be a nuisance in an icy lake, but it came in handy in your first moments of daylight. Outside the womb feels colder than inside—generally a lot colder, especially when your body is covered with fluids from the birthing process. At the same time, the newborn's body detects a rapid rise in CO_2 levels. Together these signals cause the newborn to draw a first breath, deeply. And it's a good thing the first breath is deep. It takes considerable effort because the newborn's lungs are collapsed and filled with fluid. The fluids have high surface tension, which keeps the lungs' tiny air sacs (alveoli) closed. So the first few breaths need to overcome the natural forces of surface tension and force the alveoli to open and stay open.

At this critical transition in the newborn's life, many things can go wrong. If birth is premature (less than thirty-seven weeks of gestation), before the baby's body is fully ready, the most common complication is respiratory failure. Failure to breathe early in life can be caused by problems with the respiratory center in the brain, weak respiratory muscles, or immature airways and alveoli.

To help get breathing started, the baby's lungs produce surfactant, a detergent-like molecule that reduces the surface tension in the lungs' fluids and makes it much easier for the newborn to take that first breath. Respiratory distress syndrome is caused when too little surfactant is present. This condition often affects premature babies born before thirty-five weeks, but occasionally also occurs with those born later.

Surfactant is yet another example of a seemingly modest substance playing a seemingly modest role in a much larger system—but remove this substance and the larger system threatens to break down, and at an absolutely life-critical moment.

Potential Cardiovascular Trouble in Paradise

Once the newborn's lungs are working properly, he faces yet another daunting transitional challenge. He must continue to deliver all the oxygen needed for all his cells.

Before birth, the needed circulatory flow shunts blood *away from* the lungs to the placenta, the only source of oxygen available in the womb. After birth, blood must flow *to* the lungs, as they become the only available source for oxygen. Obviously, the two circulatory shunts deployed in the womb to bypass the lungs will need to be closed. This is a major change, hitting multiple body systems at the same time. It needs to happen quickly, and permanently, else the newborn will struggle to survive.

This means certain mechanisms and orchestrations must be present to do the work. And each mechanism must be coherent—must detect when the change needs to be made, turn on the new systems at just the right moment, and turn off the old systems, also at just the right moment.

For the foramen ovale this transition is just a matter of structure and physics. Once the lungs are open and receiving most of the blood from the pulmonary artery, they oxygenate it and send it back through the pulmonary veins to the left atrium. The rapid increase in blood flow to the left atrium raises its blood pressure, so it's higher than that of the right atrium, forcing the two walls of the flap-like opening between them to come together, closing it. (See Figure 3.3.) Over the next several months this tissue fuses to permanently close the foramen ovale.

However, with modern echocardiography, it has been shown that about 25 percent of the time the foramen ovale doesn't completely close, resulting in a patent (open) foramen ovale (PFO). People with a PFO usually live normal, healthy lives. Most of them don't even know that they have a "small hole in their heart" until they undergo testing, usually for an unrelated problem.

It's possible to view this relatively high occurrence of PFO as a defect; but given the minimal risk associated with it, it can also be seen more as a normal variant. In contrast, a failure to close the second shunt (the ductus arteriosus), while very rare, is usually much more serious. If the ductus arteriosus fails to close at birth, the result quickly becomes life-threatening.

While the foramen ovale is a flap of skin that can close relatively easily, the ductus arteriosus is more akin to an open pipe, so closing it is a more complicated problem requiring a more complicated solution—an orchestrated process involving not only structure and physics, but also chemical signaling. Even though the ductus arteriosus is only for temporary use, it's nonetheless a major blood vessel with walls about as thick as the aorta and pulmonary arteries—a thickness required to support the same blood pressures. The difference is that while the aorta and pulmonary arteries never close (the result would be instant death), the ductus arteriosus must do so, and do so very soon after birth.

What kind of mechanism could solve this problem, closing just the ductus arteriosus while preventing the two associated major arteries on either side from closing? It turns out that the walls that make up the ductus arteriosus are structurally different from the walls of the aorta and pulmonary artery. Like the walls of most large arteries that handle high-pressure blood flow, the walls supporting the aorta and pulmonary artery mostly consist of elastic fibers. The walls of the ductus arteriosus consist mainly of smooth muscle.

The walls of the ductus arteriosus are structured such that it can act like an active valve, similar to the arteriolar muscles in the cardiovascular system. When the muscles of the ductus arteriosus contract enough, they can close the pipe and significantly reduce the flow—only in this case, the valve is closed just once, in the transition after birth.

The way the process works is another remarkable story. While in the womb, the higher pressure in the pulmonary artery (because the pulmonary artery is mostly closed) causes blood flow to bypass it and go into the aorta. This pressure helps keep the ductus arteriosus open. But two chemical factors also help to keep it open: low levels of O_2 and high levels of prostaglandins in the blood. Prostaglandins are lipid compounds which are produced in most fetal tissues and, notably, in the placenta.

Although the placenta provides O_2 to fetal blood, it's not as efficient as the lungs. Further, the blood coming from the umbilical vein to the right side of the fetal heart mixes with deoxygenated blood from the rest of the fetal body. Thus, O_2 levels in fetal arterial blood are only about a fourth of adult levels.

Using O_2 sensors on their surface, the smooth muscle cells in the walls of the ductus arteriosus detect low oxygen levels and respond by relaxing. They also have receptors that detect high levels of prostaglandins, causing them to relax further. So the low O_2 and high prostaglandin levels combine to keep the ductus arteriosus open in the fetus.

With birth, the baby begins to cry, the lungs open, and blood flow increases dramatically from the pulmonary artery to the billions of capillaries that line the hundreds of millions of alveoli. As blood begins flowing to the lungs, pressure in the pulmonary artery decreases, reducing blood flow through the ductus arteriosus to the aorta.

Further, opening the lungs quickly increases the O_2 level in the newborn's blood. At the same time, the newly functioning lungs begin to break down the prostaglandins in the blood. And since the placenta isn't there to produce more of them, the prostaglandin levels quickly drop. Together, these changes cause the smooth muscle around the ductus arteriosus to contract, closing the pipe.

The most difficult stages of this exquisitely orchestrated process are now complete. With the dramatically reduced blood flow through the ductus arteriosus, the cells in its walls lose their main source of O_2, which causes these cells to die. The result is permanent closure of the ductus arteriosus, usually 24 to 72 hours after birth. The ductus arteriosus is no longer needed, so the "detour" shrivels and dies, leaving just the permanent circulatory solution in place.

During an infant's life in the womb, the *ductus arteriosis* (DA) must stay open. After birth, it must close. There's little margin for error. In engineered systems, transitional stages like this are where errors are most likely to occur.

The smooth muscles of the ductus arteriosus are operated by an active control system. When open (the normal state during gestation), the channel allows blood to pass through from the side with higher pressure to the side with lower pressure. Very soon after birth, this valve must close, so the smooth muscles contract to cut off the flow.

Once again we find a control system well suited to the body's need, essential to the survival of the species, and needed just a single time, at just the right time.

Wonderment

Respiration is a tough problem for the body to solve at any time, but all the more so when there's no direct access to the air, as with a child in the womb. For respiration, a preborn baby must solve a series of hard problems, all while building out the systems he will need after birth.

As if that were not hard enough, the transition from respiration in the womb to respiration outside the womb is itself a daunting engineering challenge that requires its own set of solutions. So much needs to happen, always at the right times and never at the wrong times.

We must not gloss over the difficulty of these problems. The solutions depend on necessarily temporary systems that act as respiratory scaffolding—systems that, outside the womb, would likely cause death and therefore must be transitioned away from and discarded shortly after birth. The timing is critical. Everything must come together at just the right time, or the newborn will not survive the transition to life outside the womb.

As engineers know all too well, transitions of this kind, from one coherent solution to a different coherent solution, are fraught with the potential for errors and failures. In living systems like the human body, a transition as momentous as birth could easily lead to catastrophic failure. That such catastrophic failures occur only in a tiny fraction of births is remarkable. Why (and how) does this transition ever go right, especially given that no subsequent generations are possible if this is not achieved? It is understandable, then, why so many engineers and physicians see in all this not mindless evolution but exquisite planning and purpose.

4. Oxygen for the Body

Behold, I will cause breath to enter you, and you will live.
—Ezekiel[1]

M y (Howard's) son and his wife had just been blessed with the birth of their daughter Celina. We were especially excited because my son worked in Rome. We're Catholic, and Celina was to be baptized in St. Peter's Basilica. Family had flown in from the United States for the celebration, and the house was filled with excitement and joy.

But within a few days of her birth, Celina began to cough, wheeze, and breathe faster than normal. My son and his wife quickly realized that Celina's condition was life-threatening. The joyful mood quickly gave way to fear. What was Celina's condition, and how was it threatening her life?

I had encountered such moments countless times in my profession, but now I was facing it in an all-too-personal way: a physical breakdown underscoring just how delicately balanced the human body's sophisticated system of systems must remain in order to maintain function.

Invisible, Indispensable

At just two weeks old, Celina had come down with bronchiolitis, an infantile viral infection that constricted her airways and left her tiny body starving for oxygen. She was having so much trouble breathing that she was hospitalized and switched from room air, which contains about 21 percent oxygen, to air with a higher concentration of oxygen (around 28 percent).

Your body is made up mainly of carbon (C), hydrogen (H), nitrogen (N), and oxygen (O). These four elements combine in various arrangements to form many different types of molecules. Some of the more important molecules your body needs are molecular oxygen (O_2), water (H_2O), a sugar called glucose ($C_6H_{12}O_6$), and proteins made up of nitrogen along with the same atoms that make up sugars.

Your body consists of trillions of cells, and each must have enough energy to work properly, or it will malfunction and die. Cells get this energy from a process called cellular respiration, which breaks the chemical bonds between the atoms of the glucose molecule, releasing energy. As we saw in Chapter 1, the cell captures this energy in the ATP molecule, energy it uses to power virtually all the cell's machinery.

The chemistry of this process requires six O_2 molecules as input, so cellular respiration cannot occur unless O_2 is present. The reaction also produces carbon dioxide gas (CO_2) as a byproduct. CO_2 is toxic if allowed to build up in the body. These two chemicals must be controlled within narrow tolerances at all times, regardless of how quickly the body may change its activity level, which requires equally rapid changes in the supply of O_2 and a corresponding need to rapidly remove more CO_2.

The chemical formula for cellular respiration is as follows:

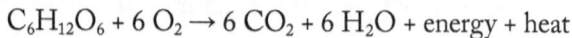

$$C_6H_{12}O_6 + 6\ O_2 \rightarrow 6\ CO_2 + 6\ H_2O + energy + heat$$

The process deploys twenty-three special enzymes in a precise series that converts one glucose molecule and six oxygen molecules into six carbon dioxide molecules and six water molecules. This process is known as aerobic respiration, meaning it uses oxygen as an input. The process releases the energy stored in the glucose molecule: 40 percent of the energy is captured by converting thirty-six low-energy adenosine diphosphate molecules (ADP) into thirty-six high-energy adenosine triphosphate (ATP) molecules. The remaining 60 percent of the energy is released as heat.

So, to stay alive your body faces a challenge. It must ensure it always has enough O_2 to meet the energy needs of its trillions of cells. And it can't allow too much CO_2 to build up.

My granddaughter Celina was at risk because her bronchiolitis was preventing her from bringing in enough O_2 and getting rid of enough CO_2 to stay alive.

The Respiratory System

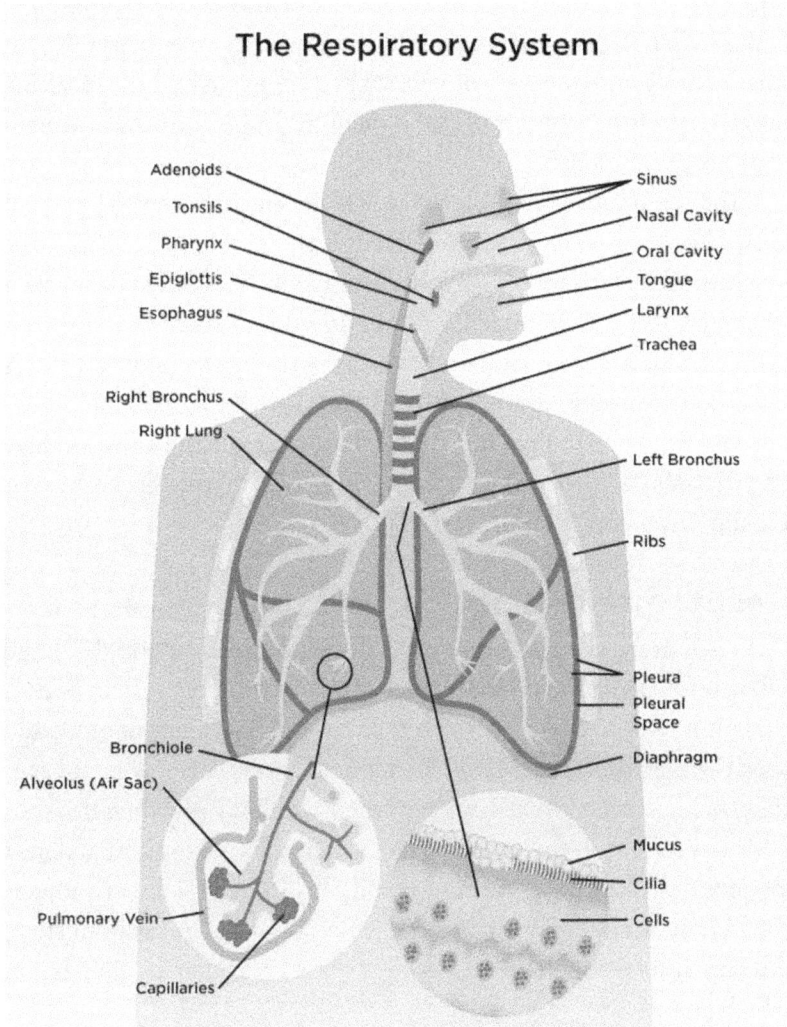

Figure 4.1. A simplified schematic of the human respiratory system.

The Respiratory System

Your body is equipped with two lungs placed within the chest cavity, which is bounded by the ribs and other bones to provide protection and support.

Your urge to breathe is generated by the cells in the respiratory center in the medulla, a part of your brainstem. The nerve messages to breathe travel through the spinal cord to the muscles of respiration— in particular, the diaphragm and the intercostal muscles between your ribs. When they contract, your chest cavity expands and the resulting negative pressure sucks air (like a vacuum) into your lungs through your nose and mouth.

The air you breathe in is about 21 percent O_2 and 0.04 percent CO_2. As you inhale, it travels through the airways, deep into the lungs, until it reaches the 600 million grapelike sacs called the alveoli. Each alveolus is surrounded by several hundred small blood vessels called capillaries. The vital exchange of gases occurs here. O_2 enters the blood from the inhaled air, and CO_2 leaves the blood and goes into the air to be exhaled. Because of this exchange, the air you exhale only has about 16 percent O_2 and a 100-fold increase in CO_2, now around 4 percent.

Control Systems at the Core of Life

The respiratory system is how your body brings in essential O_2 and offloads CO_2. But the mere presence of your respiratory system doesn't explain how your body knows to take on the right amount of O_2 and shed the right amount of CO_2. To begin to appreciate how it manages this trick, consider what happens when you hold your breath. Your body has sensors in the main arteries leading to your brain that detect the levels of O_2 and CO_2 in your blood. The CO_2 level directly affects the H^+ ion (acid) level in the blood, which is sensed in the brainstem. All this information is sent to the respiratory center in your brain, which analyzes it.

When you hold your breath, the amount of O_2 in your blood drops below normal because your cells keep using it for their energy needs even though no new O_2 is coming in. At the same time, the CO_2 level rises above normal because your cells are continually producing it, but it has nowhere to go.

When it learns of this situation from your O_2 and CO_2 (and H^+ ion) sensors, your respiratory center urges you to correct the problem by breathing. The longer you ignore the warning, the lower your O_2 drops, the higher your CO_2 rises, and the stronger the urge to give in and breathe to stay alive.

So how does this signaling system work? The body employs a triple-input control system to manage respiration. It's triggered by very low O_2 levels, or by high CO_2 or H^+ ion levels. During intense activity, the respiratory center also receives information from your muscles about their activity level and metabolism, along with insight from the higher centers of the brain about its plans. Based on all this information, the respiratory center generates the signals that tell the body how hard and fast to breathe. And it does this without any conscious effort from you.

What's required of such control systems? Even the simplest control system needs several different parts working together to achieve and maintain control. Every control system, whether in a biological or a human-engineered system, must include some means to do each of the following:

Sensing

If you can't detect what needs to be controlled, control is impossible. Sensors must be in the right places, with the right capacity and sensitivity to detect what needs to be controlled. The O_2 and CO_2 sensors are in the main arteries leading to the brain, not in, say, the spleen, where they would be of little use. And they detect what needs detecting—the blood levels of O_2 and CO_2 rather than, say, blood flow or temperature or any of a thousand other things.

Control

Control is impossible without a way to make decisions about what to do. Data from a sensor must be received, integrated with information coming in from other sensors, and interpreted. The controller must apply internal control logic to decide what to do, and then send the right signals to the right places to adjust levels as needed. The O_2 and CO_2 sensors send their information to the respiratory center in the brainstem, where the information is needed, rather than to any of a thousand other possible sites (for example, the visual or auditory cortex). The respiratory center analyzes this information and sends orders through the nerves to change the body's breathing pattern.

Effectuation

Obviously, no control is possible without some way to change what must be controlled—some organ, tissue, or other body subsystem. Control decisions are sent from the controller to the effector, where they must be received, properly interpreted, and applied. Effectors need to have the sensitivity and capacity to respond quickly enough to maintain controls within proper levels at all times. The respiratory center sends orders to the muscles of respiration, not to any of a thousand other places in the body, telling these muscles how hard and fast to breathe, which they immediately do to maintain the current level of activity.

Signaling

All control systems need some kind of signaling infrastructure to carry signals from the sensors to the controller, and from the controller to the effectors. Signals must carry the correct information, be directed to the right components, and arrive soon enough. Signaling takes many different forms in the human body.

Dynamics

The system must respond quickly enough to maintain the tight tolerances needed in the time frames needed. For example, it just wouldn't do for the oxygen control system to take ten minutes to increase oxygen levels if your body needs more oxygen right now for you to escape a raging wildfire. In the respiratory system, not only are the capacities just right, but so are the dynamics.

Harvesting and Garbage Removal

Finally, for many factors that the body must control (like O_2 and CO_2), there are two additional functions required: harvesting and garbage collection.

When a control system needs to add more of some chemical to the body, it must be able to harvest the needed raw materials from the environment—in the right amounts and at the right times—and convert them as needed for a particular use.

And when a control system needs to remove a chemical from the body, it must be able to gather that chemical and expel it into the environment—again, in the right amounts and at the right times. Any chemical, if allowed to build up in the body, will become toxic in sufficient quantities, including O_2.

Consider a home furnace. Most are fairly simple, using a single sensor designed to monitor room temperature. The control logic consists of a simple on-off switch connected to the heat sensor, which can be adjusted to the desired room temperature (what engineers call a setpoint). The effector is the furnace, which converts chemical energy (from natural gas, for example) into warm air, which is pushed into the room by a fan. When the room heats sufficiently, the sensor tells the switch logic, which turns the furnace off until the room cools and the switch logic again reactivates the furnace.

In most furnace systems, the signaling infrastructure consists of copper wires that carry the switch's status (either on or off) via a low electrical voltage to the furnace. The furnace has harvesters for both its natural gas fuel and an outside oxygen source, and a "garbage collection" subsystem to expel exhaust to the outside air.

Even the simplest furnace system has all the parts of a control system, and if you want to spend more money, an engineer can design a heating system with many more sensors, more effectors, and much more complex control logic, like you'd see in a large office building. Yet even these massive heating systems are simple compared to the controls needed for the complex distribution of O_2 and CO_2 in the human body.

From the respiratory control center in your brainstem and the bones that make up your chest cavity to your muscles of respiration, your lungs, their alveoli, the capillaries surrounding them, and the nervous system that carries these signals, it takes a lot of parts to make respiration work.

But just having the parts is not enough. The respiratory system has to be assembled, with each part in the right place, performing the right function at the right time, with the right signaling and responses in the right quantities and at the right time. Otherwise, the body will go into respiratory failure, and we know what happens then.

The human body isn't immune to death and disease, but it often proves amazingly resilient. This was the case with my granddaughter, Celina. When bronchiolitis compromised her respiration, her O_2 and CO_2 sensors informed her respiratory center of the problem, triggering it to signal her respiration muscles to correct the situation by breathing harder and faster. This enabled her body to compensate until she could obtain medical attention. After a few weeks in the hospital, Celina improved enough to go home. Her respiratory system continued to recover, and she's been a healthy little girl ever since.

The Explanatory Burden

Though we've omitted most of the complicated details, our description of respiration is medically and scientifically accurate. The facts are uncontested. Similarly, the basic engineering required to make it work is uncontroversial.

Modern science, in both biology and medicine, is pretty good at discovering and describing the body's parts, their interactions, and their processes. But it falls short in explaining how these things could have come into existence.

What, exactly, does it take to engineer all this stuff, and more importantly, to make it all work? Current explanations from evolutionists consist largely of imaginative storytelling, usually with a lot of hand-waving around the details. For instance, you might be told that as life moved from the water onto the land, the water creatures' gills evolved into the land creatures' lungs because that's what they needed to obtain oxygen from the air. That's pretty vague, and it fails to answer any of the salient questions posed by a detailed analysis of the respiratory system.

How could any unintentional evolutionary processes build the systems just described? Where did all the parts of the control systems come from? How was the orchestration for all those systems programmed? How did it manage all the engineering details we left out (which would fill several volumes)? And how could any creature survive while these things were being gradually assembled over thousands or millions of generations of evolution?

Delivering the Goods

Keep in mind, too, that this marvelously intricate respiratory system is utterly useless without the accompanying cardiovascular system for circulating oxygen-rich blood throughout the body. As an aside, since oxygen doesn't dissolve well in water, having oxygen-rich blood requires having enough red blood cells producing enough hemoglobin, which contains enough iron, to grab onto enough oxygen. (All this is explained in more detail in Chapter 6 of our book *Your Designed Body*.)

Your body is specially tuned to bring in exactly the amounts of O_2 it needs and put it into your blood. But then this oxygen-enriched blood needs to get to every one of your thirty trillion cells. That's no simple task. It involves pumping blood throughout your body by way of about 60,000 miles of pipe, your blood vessels. And just as it takes energy to roll a ball up a hill (against gravity) or push a chair across the floor (against friction) or accelerate a car (against inertia), the body must constantly use energy to work against all these forces to move blood through the blood vessels.

The heart is a muscular pump divided into right and left sides. Each side has a thin-walled upper chamber called an atrium, which acts like a holding area for incoming blood, and a more muscular lower chamber called a ventricle, which does the pumping. (See Figure 3.2.) Between the atria and the ventricles, and the ventricles and their outflow tracts, V-shaped one-way valves point in the direction of blood flow. When the valves open, they let the blood move forward in the direction it's supposed to go. And when they close, they prevent the blood from going back to where it came from.

The left ventricle pumps blood through the aortic valve into the aorta and the systemic arteries. The blood travels through progressively smaller arteries to the arterioles, and from there into the microscopic capillaries where chemicals like O_2 are offloaded to the tissues and CO_2 is taken on board. The blood then moves into the venules and through progressively larger veins on its return to the heart through the superior and inferior vena cava.

On arrival the blood enters the heart's right atrium and moves into the right ventricle. From there it's pumped through the pulmonary valve into the pulmonary arteries on its way to the lungs. In the lungs the blood travels through ever-smaller arteries and arterioles to the millions of capillaries surrounding the alveoli, where it offloads the CO_2 it picked up in the tissues and picks up a fresh supply of O_2. Then it returns to the left side of the heart through the pulmonary veins. Here it enters the left atrium, goes through the mitral valve into the left ventricle, and is again pumped into the systemic arteries.

And so continues this vital lifelong cycle: Blood coming from the lungs goes to the body, while blood coming from the body goes to the lungs. Think of it as a sort of figure-eight circulatory pattern, with the heart in the center.

The cardiovascular system (CVS) is a closed system composed of two circular pathways, or circuits. As the blood transits a complete route, it's pumped twice by the heart, once to drive it to the lungs to pick up oxygen, and a second time to drive the oxygenated blood throughout the body. Because the blood flows continuously around this closed system, O_2 delivery to the body can be adjusted simply by varying the rate the blood is pumped. It doesn't need to create more blood every time the body demands more oxygen, which is a good thing, as that would take much too long.

And it's not enough to merely move blood around the body. The cardiovascular system must adjust blood flow as the body moves and changes position. For example, after doing work at ground level for a while, when you stand up quickly the reduction in blood flowing back to your heart and from there to your brain (mainly due to gravity) can cause you to feel dizzy or even make you pass out. So your body has to act fast to ensure that your brain gets the O_2 it needs.

The cardiovascular system requires many parts that together achieve fine-grained control of blood flow:

- It has a pump and dual closed-circuit pathways to move blood around in a circle (two circles, actually). And this must happen continuously from about three weeks after conception until you die.

- It uses special unidirectional valves in the heart to prevent backflow and force the blood to always move in the right direction. When the blood pressure on the upstream side is higher than on the downstream side, the valve opens and allows blood to flow through. When the pressure reverses (when the ventricle pump relaxes), the valve closes and prevents blood from flowing in the wrong direction. These valves must have precise shapes and strengths for this purpose, and point in the right direction.

- It uses multiple control systems to achieve the fine-grained flow adjustments needed to direct blood in the right quantities to exactly the right areas at the right times. The systems include adjustable pressure, individually adjustable valves, and controls localized to specific muscle groups.

- It uses multiple signaling pathways to pass requisite information across all the varied pieces and parts scattered throughout the body.

The cardiovascular system is thus able to react to the body's changing needs for oxygen, usually even faster than the body's needs change. It's like a pipe organ, in which a single pump generates air pressure and the many pipes each have a valve. When the organist presses key combinations on the keyboard, valves at specific pipes open (some more, some less) to achieve the correct timbre, and music happens.

The system's plumbing architecture is ingenious. Closest to the heart the arteries are larger, with less resistance, the better to handle higher and faster flows. Closest to the heart the arteries have thicker walls to handle the higher pressures. This distributes oxygenated blood to remote areas of the body swiftly. The arteries branch and become progressively smaller the closer they get to their destination. As the pipes get smaller, their resistance increases, and the blood pressure and flow rate decrease. With lower blood pressure, the vessel walls can be thinner.

When the blood reaches the capillaries, where O_2 and CO_2 is exchanged, the vessel walls are extremely thin and the blood has slowed to a crawl. Here the blood vessels are so small that individual red blood cells can only pass through one at a time, and then only by being squeezed as they pass through. The squeezing action deforms and elongates the red blood cells, increasing the contact area between the blood cell and the extremely thin vessel walls. The thinner capillary walls, the slow blood flow, and the red blood cell deformation caused by the narrow capillary diameter all combine to maximize the efficiency of the exchange of O_2, CO_2, and pretty much everything else the cells need.

The cardiovascular system is a wonder. It is an extraordinarily complex system of systems in which each part must be present, all the parts must be properly connected, and each part must function correctly—with precise timing, precise signaling, precise controls, and precise plumbing—to solve the set of difficult problems imposed on the body by the laws of physics and chemistry.

To work, the cardiovascular system must be more than complex; it must be coherent. All must be just right or else the system doesn't work, and when the system doesn't work, the body doesn't work. Moreover, all the orchestrated precision that causes the system to work happens subconsciously, so we don't have to spend all day thinking about how to balance our blood flow.

As a whole, and in each of the parts, the cardiovascular system presents an enormous explanatory challenge to any theory purporting to explain its origin.

5. Seeing Is Believing

How came the bodies of animals to be contrived with so
much art, and for what ends were their several parts?
Was the eye contrived without skill in opticks?
—Isaac Newton[1]

A NTHONY IS A HEAVY EQUIPMENT OPERATOR, MARRIED AND with children in high school. One day he notices that sometimes when he shifts lanes in traffic, a car seems to appear out of nowhere. Another day at work, while running a front-end loader, he has a near miss when he doesn't see a foreman walking nearby. The near miss shakes up both him and the foreman, and he's ordered to get himself checked over right away.

After a full assessment, his eye doctor explains that, while his near and far vision are perfect, his peripheral vision has degraded. Anthony can see things in front of him clearly. It's the things coming at him from the side that give him trouble.

The doctor tells him his condition is serious, possibly permanent, and isn't correctable with glasses. Anthony fears he may never again be able to work as a heavy equipment operator, or even drive. His wife drives him home. Along the way he wonders aloud why they can't fix his problem. "If we can put a man on the moon," he says, "they should be able to fix me!"

What Anthony doesn't realize is that the human vision system is far more complex than the advanced engineering required for the moon landing.

Seeing Is Hard to Do

It's often said that the eye is like a camera. But this is backwards, because the vertebrate eye was around long before the first camera. Better to say that a camera is like an eye—albeit an eye in a much simpler form.

It turns out that sight is yet another hard problem. Light needs to come into the body, be converted into signals, and then be sent to the brain to be processed. And that's putting it simplistically. Vision requires many millions of light sensors sending data to the brain, which requires massive parallel processing to convert the signals into a view of the surrounding world. All this must happen in real time, with no perceptible delay between something happening and the experience of seeing it happen. A lot of stuff will need to be "just right." For this to happen, a series of problems must all be solved at the same time for the vision system to work.

With a little engineering thought, we can predict much of what will be needed in order to detect and use the natural phenomenon of light:

- Sensory transducers to convert light into information that can be interpreted.

- Numerous individual sensors, precisely arranged.

- Signaling of sufficient quality and quantity (bandwidth) to send information from the sensor array to the brain.

- Substantial processing power to convert the raw signals into a consistent, whole, and timely image.

- A structure able to yoke all the parts in a precise configuration while enabling them to move or rotate as needed for their functions.

- A finely tuned optical system that can quickly adjust to different situations to provide a sharp image in a wide variety of light intensities and across a variety of frequencies.

Eye Anatomy

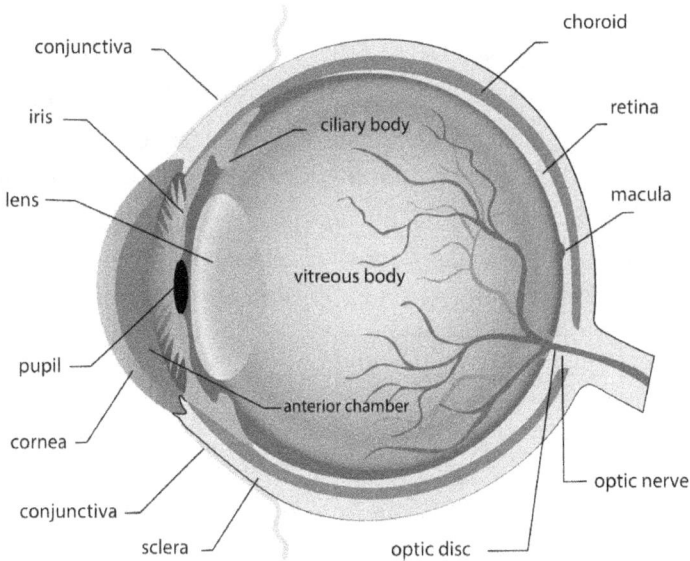

Figure 5.1. A cutaway view showing the main parts of the eye.

The most critical requirement for vision is some kind of sensor. Just as a camera's sensor (or its film) converts light into information, the eye's sensors need to convert individual photons of light into electrical signals. Figuring out how to accomplish this is a challenge demanding enough to get an engineer's juices flowing.

The retina is where this problem is solved. (See Figure 5.1.) A thin layer of tissue along the back of the retina contains two specialized types of photoreceptor cells called rods and cones because they resemble these shapes.

Each of your eyes has about 120 million rod cells in its retina, mostly scattered throughout the outer edges. Rod cells make a light-sensitive molecule called rhodopsin, which reacts to any wavelength in the visible spectrum. Because they're not picky about specific wavelengths, these molecules mainly detect luminosity rather than color. In effect, they see in black and white. They're mostly located off the central axis of vision, at the periphery, providing peripheral vision.

For central vision, each of your eyes has about six million cone cells. Each cone makes one of three light-sensitive molecules, together called photopsins, which react to specific wavelengths of visible light—either red, green, or blue. The highest concentration of color-sensitive cones is right in the middle of the eye's optical axis—the macula and, in particular, the fovea. Here is where the most and best information is needed. As a result of this concentration, the sharpest acuity with the best color fidelity is available exactly along the central axis of vision—right where your eye is aimed.

How do these photoreceptors work? It takes some special machinery to convert photons of light (traveling at light speed) into something the body can make sense of.

STRUCTURE OF THE RETINA

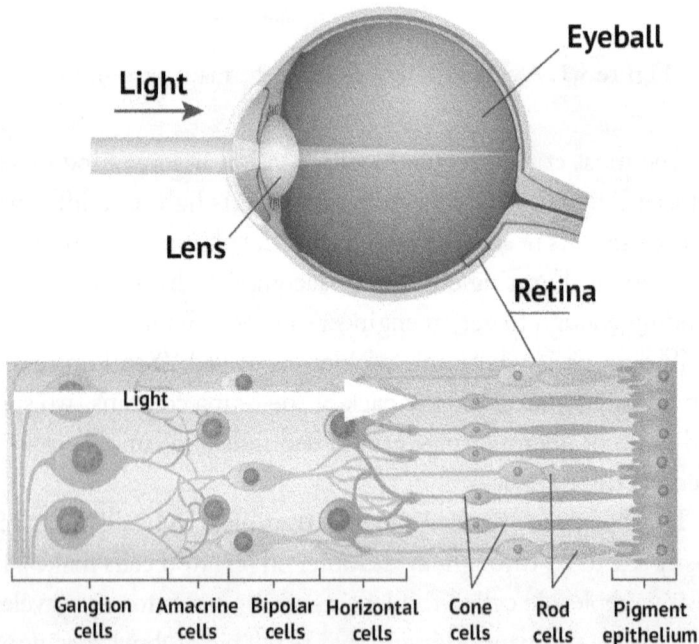

Figure 5.2. The upper image shows the location of the retina in the human eye. The lower image provides details of the retina's structure.

When a photon collides with a photosensitive rhodopsin molecule (in a rod cell), the energy changes the molecule's shape, which triggers the photoreceptor to alter its signal to a nearby interconnecting nerve cell. These auxiliary nerve cells (mostly bipolar and ganglion cells) collect the visual sensory information from the nearby photoreceptor cells, preprocess it, and ultimately send the results along the optic nerve to the brain. The photopsin molecules in cone cells work in a similar way.

What follows is a summary of the biomolecular process at play.[2] When a photon of light enters a rod cell, the smaller 11-*cis*-retinal molecule attached to the larger rhodopsin protein changes shape to become 11-*trans*-retinal, a transition that occurs in mere pico-seconds.[3] (A pico-second is one trillionth of a second.) This change in turn causes rhodopsin to change its shape as well. This starts a cascade of several chemical reactions involving several different molecules, a cascade that ultimately results in the reduction of an important signaling molecule in the cell called cGMP. The sudden drop in cGMP stops the rod cell from sending out its neurotransmitter (glutamate), which normally inhibits the nearby bipolar cell. With the inhibitor removed, the bipolar cell sends an electrical signal to the ganglion cell, which then sends a signal along the optic nerve to the brain. Speed is essential if we are to experience vision as instantaneous. And it's crucial that we do. In most situations, it wouldn't do much good if you didn't see what was happening until some while after it occurred.

This is a lot of action, all of which needs to be just right or vision will not happen.

Curious readers might be wondering where all these precision parts come from and how they come together. What raw materials are required, how much information in the DNA is required, and how is the whole assembled and activated, and only in the exact cells where it's needed for vision? Good questions, all. But here notice that this is an example of a recurring design pattern in the human body, what we term the *push-pull principle*: Two separate systems are required to achieve even the most basic coherent functions. One to ramp things up, another to ramp things down. One to start, another to stop. One to send, another to receive. One to push, another to pull.

In this case, we just saw that we need one just-right chemical cascade to turn a photon into a neural signal. But once a given photosensitive molecule has changed its shape in response to an incoming photon, that molecule can't react to another photon until it has been reset to its original shape. It's like a spring-loaded animal trap that needs to be reset before it can spring again.

It turns out that a completely separate just-right cascade is needed to turn the photosensitive molecule back into its original shape. Resetting the shape of the retinal (and subsequently the rhodopsin) is not simple. It's another example of a thorny problem created by an ingenious solution, with the new problem itself requiring another, different, ingenious solution. In this case, three distinct tailor-made enzymes, working in perfect sequence, are required to convert a photosensitive molecule (rhodopsin) from its spent shape back to its ready shape. And this doesn't happen in the photoreceptor cells. It happens in a special tissue called the retinal pigment epithelium (RPE), which sits right behind the photoreceptor cell.

In yet another stage in this series of solve-or-die problems, the photosensitive molecules in the rods and cones are built on vitamin A. So, if your body couldn't get vitamin A, you wouldn't have rhodopsin or the photopsins, your rod and cone cells would be useless, and you'd be blind. We won't unpack the many things required to harvest, transport, and process the vitamin A. Suffice to say, it takes several steps to absorb vitamin A in the intestine, multiple kinds of transport proteins to move it through the body, and still other protein types, including various enzymes, to use it to build one of the photosensitive molecules needed for vision. So, we need yet another entirely separate subsystem of mechanisms and controls to make this happen.

Fuel for the Fire

The process of seeing light uses a lot of energy. In fact, the cells in the retina have the highest metabolic rate in the human body.

This means they need a steady and generous supply of O_2 and other nutrients (including vitamin A), and they need to get rid of plenty of CO_2 and other waste products. For this, they need a lot of blood circulating nearby.

To support the energy-hungry retina, the eye has an especially dense network of capillaries in a special layer of tissue called the choroid, just behind the retinal pigment epithelium in the retina (in the back layer of the retina). These capillaries deliver just the right mixture of nutrients and remove enough waste products to support the energy needs of vision—full-time, all day long.

Poor Design?

Despite the coherent design and workings of the eye's many well-coordinated subsystems, in recent years the eye has become a folk-legend for Darwinist storytellers, who have declared it to be an example of the "poor design" that results from eons of purposeless accidents driven by natural selection. Their claim goes something like this: *The photoreceptor cells in the eye point away from the front of the eye, where the light comes from. This is obviously backwards. The auxiliary nerve cells that preprocess the visual signals are situated between the front of the eye and the photoreceptor cells, which surely compromises vision at least somewhat. This is bad design, best explained not by reference to some all-wise designer but as the outcome of the aimless creative process of evolution by random mutation and natural selection.*

Stories of this kind can't withstand much scrutiny, especially scrutiny from people who understand the practical necessities of real systems. Due to the energy-hungry metabolism required for human vision and the need to constantly reset all the photosensitive molecules, the rod and cone cells must be very close to the retinal pigment epithelium, which performs both these tasks. So, in this case, "backwards" is best—clever rather than poor design.

The Optics of It

There's more to vision than merely processing light at the back of your eyes. If it's going to be of much use, the light must get there in the first place and be properly focused. And this leads us to the next set of problems the body must solve: the problems of classical optics.

The cornea is a convex (curved outward) transparent connective tissue made from a special type of collagen. It protects the front of the eye while allowing light to enter. Since light rays coming into the eye are either practically parallel (if the viewed object is more than about twenty feet away) or diverging (if the viewed object is closer than twenty feet), they must be bent (refracted) so that they all come to a focal point on the retina, preferably at the fovea, which has the highest density of cone cells. The curvature of the cornea provides 60–70 percent of the refractive power needed to achieve focus.

Blood cells are opaque, so if the cornea had blood vessels it wouldn't be transparent. This presents a problem for the cells in the cornea, which still need a way to get O_2, water, and nutrients. So instead the cornea is "fed" from two sources: the tears that wash across the front of it by the eyelids when they blink, and the clear fluid (aqueous humor) in the chamber that sits between the cornea and the lens.

The lens, also transparent, is an elastic biconvex (doubly outward curved) structure consisting of connective tissue containing collagen. Like the cornea, it lacks blood vessels and gets O_2, water, and nutrients from the transparent aqueous humor.

Indeed, all the tissues and fluids between the front exterior of the eye and the retina must be transparent and colorless. Vision would be impossible otherwise. Consider how difficult that is to achieve. The eye contains the only tissue in the body that is truly transparent and colorless, and it's located exactly where it's essential for function, and nowhere else.

Once again, the right materials for the right uses, in the right places.

The lens is suspended in place by a ring of ligaments attached to a ring of ciliary muscle. When the muscle contracts, the lens bulges to enable closer focus. When the ciliary muscle relaxes, it flattens the lens for more distant vision. This action dynamically fine-tunes the focus in an automatic control process called the accommodation reflex.

Depending on how far away the object of interest is, the lens provides 30–40 percent of the refractive power needed to achieve focus.

The eye's cornea and lens work exactly like a two-element lens on a modern camera, though their optical formula is simpler because the eye's sensor is curved while most photographic cameras have a flat sensor, which is optically harder to focus on. But the way the eye lens achieves variable fine-tuned focus is unlike that of any modern-day camera. It's better, so much so that mimicking the eye's lens would be a huge breakthrough in human-designed optics.

There's one more optical problem our eyes need to solve, and solve really well: modulating how much light enters the eye. When it's dark, it's hard to see. And when it's too bright, it's harder still. Happily, our vision system is up to the challenge.

The choroid layer of tissue behind the retina curves around the eye to the front, and at the front (and perfectly aligned with the optical axis of the eye), the choroid has an opening, called the pupil, surrounded by the colored iris (which is not transparent and blocks light). The iris is operated by two different muscles. These muscles are controlled by the pupillary light reflex, which automatically controls the amount of light entering the eye.

Enter a dark room and the dilating muscle of the iris enlarges the pupil to let in more light. Shine a bright light into the eye and the contracting muscle of the iris tightens to shrink the pupil. You can see this reflex in action if you look in the mirror and open and close your eyes or shine a light in them.

When light is painfully bright, another reflex automatically closes the eyelids as a further layer of protection for the sensitive retinal tissues. And if it's really bright out but you still need to see, you reflexively squint, using your eyelids to further reduce the amount of light that enters your eyes.

Finally, when it's dark, as at dusk, your cones are too light-hungry to function well. But this is where the rods come into their own. Rods, recall, can't detect color like cones can, but rods gain something in the trade-off. They are much more sensitive to light than the cones are, making them ideal for vision in poor light. (Rods also are better at detecting motion.)

The pupillary reflex is yet another example of an autonomous control system essential for good vision. It provides a critical function to the eye and a protective reflex that prevents damage. This ingenious feature has been copied in cameras, albeit crudely. Such copying is an example of what is known as biomimetics, where human inventors borrow clever engineering solutions from the biological world. Other examples of inventors getting ideas from nature include airplanes, hook-and-loop fasteners, suction cups, sonar, the light bulb, and tape. And of course, many mechanisms in the field of robotics.

The Supporting Cast

Just as the lens and image sensor of a camera need the other parts of the camera, your eye needs a host of other systems and structures to work properly.

Five bones, each tailored to its role, make up the orbital cavity in the skull. These protect about two-thirds of the eyeball and provide the base for the origin tendons of the six muscles that rotate the eye in its socket. The eyelids and eyelashes protect the eye from dust, dirt, bacteria, and other foreign objects. A film of tears consisting of oil, water, and mucus is produced by the oil glands of the eyelids, the lacrimal gland, and the mucus membrane that rests on the sclera (the white outer protective coating of the eyeball). The tear film lubricates the eye, protects it from infection and injury, nourishes the surrounding tissue, and preserves a smooth surface to aid in optical focusing.

The vitreous humor is a thick, transparent, gelatinous substance that forms and shapes the eyeball. It can be compressed and then return to its natural position, allowing the eyeball to withstand most common physical stresses without serious injury.

The corneal reflex (blinking) is a protective reflex brought on by something touching the cornea. The menacing reflex, something coming at you quickly, and the optical reflex, sudden exposure to bright light, cause the blink reflex as well. The latter two depend on visual sensory information sent to the brainstem.

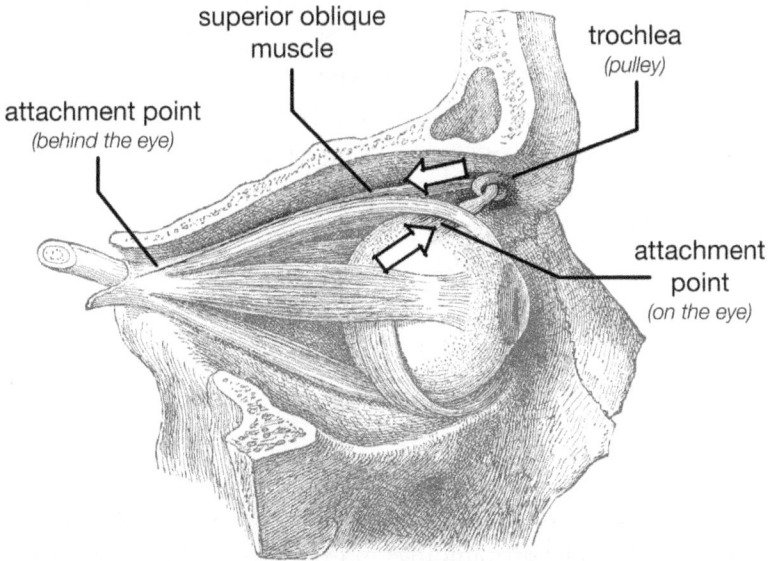

Figure 5.3. The trochlea in the right eye, viewed from the side. The arrows indicate the direction of the pulling force when the superior oblique muscle contracts to rotate the eye. The trochlea acts like a pulley to redirect the force.

The trochlea is a good example of a thorny assembly problem and its solution. (See Figure 5.3.) The eye is fixed in a bony socket, with six muscles attached. Each muscle rotates the eye in a different direction. But there's a problem. Due to the shape of the skull and the positioning of the eyes, there's no suitable attachment point for one of the six muscles. So the eye employs a clever pulley system to redirect the energy of one of the muscles. "Trochlea" is Latin for "pulley," and as you can see in the illustration, this is how it works.

Assembling this structure involves something like "threading the needle." How is a structure like this made during fetal development? How are the assembly instructions encoded?

Like every other part in your body, the eyes can't just be floating around. It also wouldn't do for them to roll out of your head and fall to the floor whenever you turned your head quickly. They need to be anchored, and they need to be held in the right configuration, with specialized structures surrounding them and protecting them, while at the same time being directly exposed to the light outside the body. As if this weren't challenge enough, the eyes also must be allowed to rotate up and down and side to side.

For human eyes, this involves five specially shaped bones forming a ball-and-socket structure, with six muscles for each eye, with their tendons attached at just the right points to the bones. Each of these parts must be correctly specialized, organized, and assembled. This is coherence of a high order.

One of the keys to vision is the ability to track a moving subject. And once again, it takes some interesting mechanisms to make this work.

The vestibular-ocular (doll's eyes) reflex helps stabilize the retinal image. When you're in motion, unless you focus your view on something specific, your surroundings move across your visual field at the same speed as you do. Without the ability to control your eye movements when your head moves, what you look at would be blurry. Look into a mirror and focus on your eyes as you rotate your head from side to side and up and down. Notice how your eyes automatically move in the opposite direction of your head so you can keep your subject in focus and in the center of your visual field.

The brain combines sensory information from the eyes with angular information about head motion from the semicircular canals in the ears to reflexively move the eyes to maintain a stable retinal image. This means we can visually track objects no matter how fast or in what direction they (or we) move.

Even with all the other visual mechanisms discussed above—the sensors, biochemistry, optics, and so on—we'd be functionally blind without this intriguing ability to focus on a subject and track it as we (and/or it) move.

Optical Pathways

All the above are necessary if we're to see the world around us. But we still haven't explored how the visual information from all those rods and cones gets to the brain, or how the brain turns those signals into a unified and complete visual image.

About 20 percent of optic nerve impulses veer off early to provide sensory data to the brainstem, which controls the various reflexes discussed above. Because these go to the brainstem, we can diagnose brainstem function using the reflexes in the eyes. If the brainstem is working right, touching the cornea with a wisp of tissue should make the eyes blink. Shining a light into either eye should make the pupils contract. And twisting the head from side to side should make the eyes move in the opposite direction the head is moving.

What about the other 80 percent of the signals from the optic nerves? These go to the visual cortex, but they don't take the path you might expect.

Just as a camera lens forms an inverted image on the sensor, so too does the retina. What you see in the upper part of your field of view is sensed in the lower part of your retina, and what you see to your left will be sensed in the right side of your retina. This means the image is backwards and upside-down on the retina. This is just simple optics, but the brain's vision-processing system needs to properly decipher these signals so that you sense the world as right-side-up and skillfully navigate through it.

There's another complicating factor. Your vision is split vertically in each eye, and each half of the visual signal—the inside (nasal) and the outside (temporal)—takes a different path to the visual cortex.

VISUAL PATHWAY

top view

LEFT VISUAL FIELD ⋮ RIGHT VISUAL FIELD

binocular field

visual field
of left eye

visual field
of right eye

optic nerve

pituitary gland

optic chiasm

optic tract

suprachiasmatic
nucleus of
hypothalamus

lateral
geniculate
nucleus
of thalamus

left visual
cortex

right visual
cortex

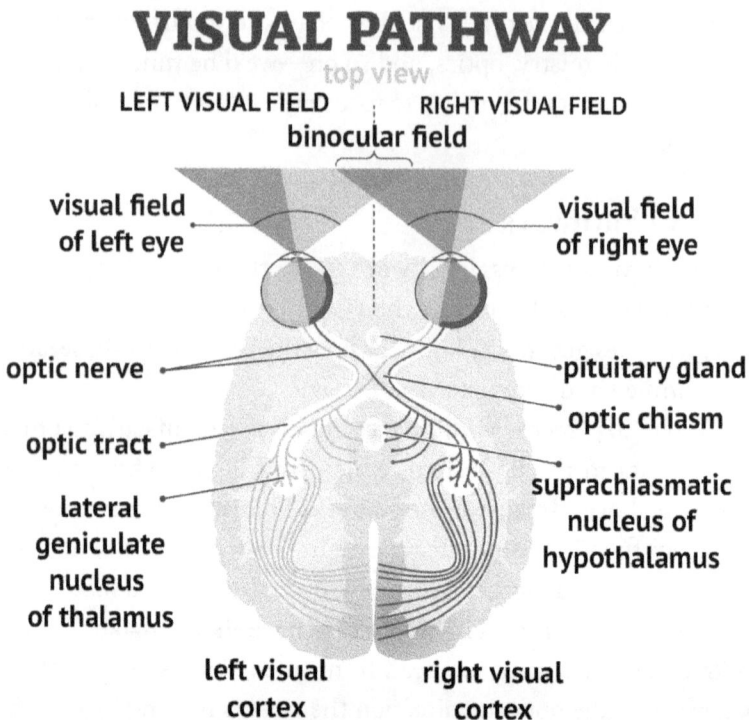

Figure 5.4. The pathway for visual information from the retina to the visual cortex.

Imagine a vertical line going through the middle of the retina. All the photoreceptors to the right in both eyes—the inside (nasal) half from the left eye and the outside (temporal) half from the right eye—send their messages to the right side of the brain. (See Figure 5.4.) Similarly, the photoreceptors to the left in both eyes—the outside (temporal) half from the left eye and the inside (nasal) half from the right eye—send their messages to the left side of the brain. But because the image on the retina is inverted, the right visual cortex processes everything in the left visual field of each eye and the left visual cortex processes everything in the right visual field of each eye. The information pathways cross over and go their separate ways at a point called the optic chiasm.

There's one more complication. Each eye sees from a slightly different perspective, a property that makes it more challenging to assemble a complete visual image. But the complication is worth it, and not just because having two eyes provides redundancy (lose one eye and you can still see). Having two eyes also allows you to sense distances—depth perception. That each eye sees an object from a slightly different angle enables the brain to calculate the resulting differences and accurately determine how far away various objects are. From this, your brain builds a real-time, 3-D model of the world around you, which comes in handy for spearing a fish moving through a stream, hitting a moving target with an arrow, or catching a fly ball in center field.

Putting It All Together

Now it's time for the brain to do its job. It must take this conglomeration of photon-generated nerve impulses—turned around, upside down, split-up, overlapping, with a blend of color and black-and-white info, and with two holes in it (from the blind spots in each eye, where the optic nerve exits the retina)—and turn it into the unified whole we experience as vision. And it must do this so fast that we can't detect a lag between the experience in the world and our vision of it.

Though we can describe many parts of the overall vision process, no one really understands exactly how it works. How, exactly, does our mind make sense of the jumble of data it receives from the many millions of photoreceptors from two eyes? It seems that the more we learn, the more profound the mysteries we discover.

Since each rod or cone in the eye produces its own signal, it acts a bit like a single pixel in a digital camera sensor. It's imperative that the signaling and interpretation system(s) keep track of (know?) where each signal is coming from and how that pixel relates to the others around it. It just wouldn't do if the pixels arrived in random order and could not be reassembled in the right order. How does the brain know where each signal is coming from? How does it assemble all these signals across different pathways into a coherent visual whole? The task is so outlandishly complicated that scientists have yet to unravel how your body manages it.

What's Wrong with Anthony's Vision?

The above survey of human vision is, keep in mind, a massively simpli-fied overview of a far more complicated system of systems. But with this basic understanding of visual signaling pathways, and one or two additional details, we have enough to get a handle on what is messing up the peripheral vision of Anthony the trucker.

An untrained person might guess that Anthony has a defect in his cornea or lens, an aberration along the sides that obstructs peripheral vision. But, no, everything in his eyes looks normal. The problem lies elsewhere, and the eye doctor quickly arrives at a provisional (and clas-sic) diagnosis: Given the different pathways the visual signals travel, the most likely location of Anthony's problem is where the signals for the outer (temporal) field of vision for both eyes run together but where the signals for the inner (nasal) field of vision (for both eyes) may not be affected. This only happens in one place in the body—the optic chiasm, where the two information pathways from your two eyes cross before going their separate ways.

The eye doctor soon pinpoints Anthony's precise condition: bi-lateral loss of peripheral vision likely due to a pituitary tumor. That's because the pituitary gland sits just below the optic chiasm. A tumor there can press on the nearby nerves, rendering them non-functional. (Nerves don't work well under physical pressure, like when you hit your funny bone.)

Anthony's eye doctor orders a brain MRI, which verifies his suspicion. A neurosurgeon removes a large benign (non-cancerous) pituitary tumor. Eventually Anthony's peripheral vision returns to normal, and he is able to go back to work.

A Chemically Fine-Tuned Biomechanical Electro-Optical Signal-Processing Vision-Interpretation System

Our vision requires perfectly tuned physical structures with orches-trated motion, transparent tissues and fluids (in all the right places and in none of the wrong ones), unique supply chain solutions, con-stant delivery of energy to energy-hungry cells, layers of complex control systems, high information signaling systems, and information processing and image assembly on a scale we've barely begun to understand. And all this with no perceptible delay and no conscious effort so you can focus on what you're seeing rather than how you're seeing.

Vision is both a wonder and an enigma.

The story of Anthony illustrates how critical the signaling pathways in the body are. Even if all the other components of vision are present and working properly, a limit on the signals from the eyes to the brain can significantly diminish eyesight. It's another case of the interwoven dependencies among the body's systems. Dependencies like this are common to highly complex human-engineered systems, suggesting that they are unavoidable for systems this sophisticated. This brings us back to one of this book's central questions for evolutionary theory: Could an organ system of such sophistication, necessarily involving so many layers of dependencies, have evolved in a series of incremental and functional steps? Does a continuously functional pathway (an adaptive continuum) exist for such an evolutionary journey? If so, what is it?

Darwin himself wondered whether the eye could ever be explained by the mechanisms of his theory. He confessed that the idea of the mammalian eye having arisen via natural selection "seems absurd in the highest possible degree."[4] But still he held out hope that an adaptive continuum would one day be found to explain the gradual appearance of so perfect an organ.

Of course, he and everyone else at the time had a hopelessly simplistic understanding of the physics and mechanics of vision, including the molecular chemistry and visual processing involved. We have discovered so much since then. As a result, we understand the challenges facing the evolution of vision far better than Darwin did.

Even today our understanding of the ins and outs of vision is grossly incomplete. Nevertheless, we can safely infer that as our understanding of the vision system grows, we will uncover (and be amazed by) additional clever engineering solutions, all working together, and each such solution posing a fresh challenge to evolution.

6. Hearing Is Believing

The design of the human ear is one of nature's engineering
marvels.... It is not only a perfect design but also a
low–cost design.... In terms of performance, the human
ear design is very impressive and serves as an inspiration
for designing products in industrial environments.
—Sundar, Chowdhury, and Kamarthi[1]

To hear, the body must collect acoustic signals from the environment (pressure waves in the air), channel them to the right locations, convert them into nerve impulses, send them to the brain, and correctly interpret them into experiences like speech and music. And, just as with vision, if any one of those parts works incorrectly, or even just a bit less efficiently, hearing is either severely degraded or impossible.

The human ear can detect sound when the eardrum is displaced by as little as one-tenth the diameter of a single hydrogen atom. Yet it can also hear and correctly interpret sounds with acoustic pressure levels approaching the loudest sounds produced in nature (~1 kilopascal [kPa]).

And you can do more than register sounds of varying pitch and volume. From an early age you could tell from the sound of your mom's voice just how much trouble you were in, and which direction she was calling from (so you knew which way to run). These and other features of human hearing require—and by now this should come as no surprise to readers—not just one or two clever engineering solutions, but a host of ingenious solutions upon ingenious solutions.

Figure 6.1 illustrates the main parts of the body's auditory system. Its many parts work together to gather sound waves from the environment and transmit them accurately and efficiently to the cochlea, where a subsystem called the organ of Corti converts them into nerve impulses and sends them to the brain.

ANATOMY OF THE EAR

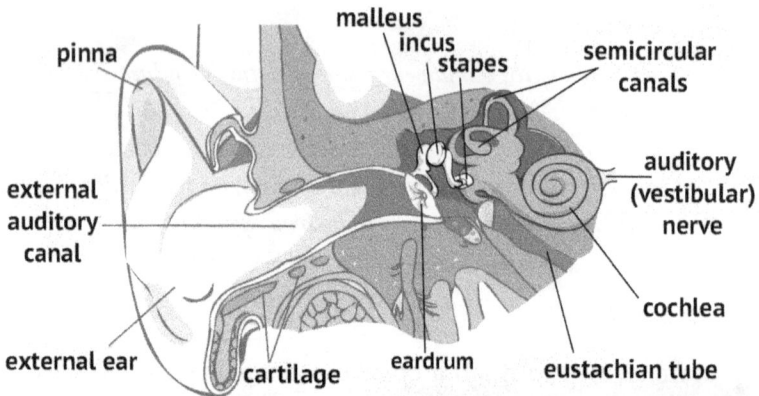

Figure 6.1. The parts of the ear.

The ear is divided into three regions: the outer (external) ear, the middle ear, and the inner ear. We'll walk through these parts in order—that is, following a sound wave as it moves from outside the body to the inside where it's converted into information and then into an experience.

The Outer Ear

The outer ear is made up of the pinna (ear flap), the ear canal, and the tympanic membrane (eardrum).

The pinna acts like a satellite dish, collecting sound waves and funneling them down the ear canal. But it does more than just collect. The pinna's ridges and folds reflect and absorb certain frequency components of incoming sound waves. Since the pinna is not circularly symmetric, sounds coming from different directions have slightly different acoustic characteristics. This means certain frequencies in a sound will be slightly softer or louder depending on the direction they enter the ear. This allows you to tell the direction a sound comes from. This is why we instinctively look up when we hear a sound coming from above us.

To further help with this, we have two ears for stereo sound. We can detect differences as small as ten microseconds in the arrival time of the same sound in each ear. We can also detect subtle differences in loudness between our two ears. Coupled with the fine-grained sound-shaping done by the outer ear, this allows us to tell the direction of a noise and hear in three dimensions. That is, our minds can generate a three-dimensional understanding of what's going on around us based solely on sounds.

Close your eyes and listen carefully to the sounds you hear. Where are they, both in direction (left or right, front or back, up or down) and distance away from you?

The ear canal is a hollow tube about two centimeters long. It forms an acoustic channel between the pinna and the eardrum. The ear canal may not seem interesting at first glance, but its length plays a crucial role in hearing.

Much like a pipe in a pipe organ, the outer ear consists of a rigid tube open at one end and sealed at the other. Incoming waves bounce off the closed end and create standing waves in the tube (ear canal). This amplifies sounds at or near the tube's resonant frequencies (constructive interference) and dampens sounds at other frequencies (destructive interference). This increases sensitivity to particular frequencies. Basically, it's a passive amplifier!

For the human ear, this amplification is strongest at around 3,000 Hz. While this is higher than the central frequencies of human speech, it's exactly the range where the percussive elements of the consonants in human speech are most prominent, and the consonants are essential for distinguishing the nuances of human speech.

The net effect is that the outer ear preprocesses incoming sound waves to maximize sensitivity to the natural frequencies of human speech. That is, our ears are fine-tuned to hear best at the same frequencies we naturally speak.

The human ear can hear sounds from 20 Hz to around 20,000 Hz. Normal human speech ranges from 80 to 2500 Hz. The lowest note on a subcontrabass tuba is at or slightly below the lower limit for human hearing, middle C on a piano is 262 Hz, and the highest note on a flute is 2,093 Hz.

The eardrum (tympanum) is a small membrane, about one centimeter in diameter, at the inner end of the ear canal. It's a durable piece of skin tightly stretched across an opening in the bony skull. The eardrum vibrates at the same frequency as an incoming sound wave, enabling it to accurately and efficiently transmit sounds from outside the body to the inside. All the while, it maintains a barrier that seals the delicate inner workings of the ear from foreign matter and bacteria.

> The eardrum is perfectly engineered to perform two distinct and opposing tasks. First, it must provide a barrier to prevent harmful contaminants from reaching the inner ear. For this it must be durable and fairly rugged. Second, it must be thin and lightweight to transmit sounds from outside the ear to inside the ear, with acoustic accuracy and without appreciable degradation or damping of the sound's energy.

The Middle Ear

The middle ear is an enclosed air-filled chamber, beginning at the inner surface of the eardrum and ending at the cochlea.

The middle ear contains the *ossicles*, the three smallest bones in the body. These are the malleus (hammer), incus (anvil), and stapes (stirrup). They were given their familiar names because they resemble those objects in shape. Working together, they transmit the vibrations of the eardrum into the inner ear.

To do this, the malleus is attached to the eardrum and the incus, the incus to the malleus and the stapes, and the stapes to the incus and the oval window of the cochlea, as shown in Figure 6.2.

THE MIDDLE EAR

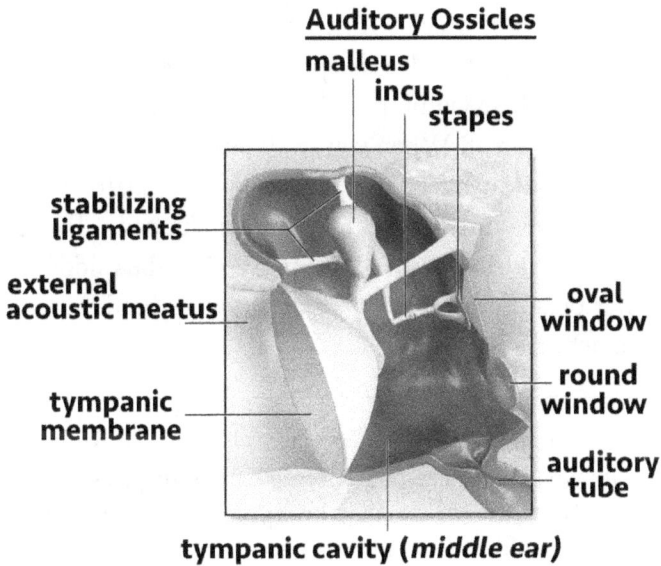

Auditory Ossicles
malleus
incus
stapes

stabilizing ligaments

external acoustic meatus

tympanic membrane

oval window

round window

auditory tube

tympanic cavity (*middle ear*)

Figure 6.2. The parts of the middle ear. The oval and round windows are parts of the cochlea.

Sound waves make the eardrum vibrate, which vibrates the malleus, which vibrates the incus, which vibrates the stapes, which vibrates the oval window of the cochlea. But the key to hearing is how these bones are precisely shaped and interconnected to modify incoming vibrations. We'll look at this below.

Interestingly, these bones are fully formed at birth and do not grow as the entire body around them grows from infancy to adulthood. These are the only bones in the body with this property. How does the body grow all its other bones while keeping just these ones from growing? What mechanisms and control systems are needed? So far, neither medical science nor biology has answers, but engineers know that such things don't happen by accident, so there seem to be many interesting discoveries yet to be made.

As you'd expect by now, there are some less-than-obvious problems with hearing that the body needs to solve.

For instance, the middle ear is filled with air, but without an innovation the tissue in direct contact with the air would gradually absorb it, causing a vacuum effect, which would reduce eardrum movement and impair hearing. To solve this problem, the ear uses a small tube, called the eustachian (auditory) tube, that connects the middle ear to the back of the throat. When you swallow or yawn, this tube opens, allowing fresh air to enter the middle ear. This equalizes the middle ear's air pressure with the pressure outside the body. This tube can get clogged, as during a head cold, preventing the middle ear from equalizing pressure, which, as we all know, degrades hearing and causes earaches.

Sound is composed of pressure waves. For the ear to be able to accurately register these pressure waves, the air pressures on the two sides of the eardrum must be equal. When the air pressure between the middle ear and the surrounding environment is different, your hearing must fight against this difference. This affects both the amplitude (loudness) and the tonality (pitch) of incoming sounds.

As a second and more formidable problem, sounds entering the body come through the air, but the cochlea is filled with fluid. The cochlea's fluid, as we'll see, serves a vital purpose, but it presents a thorny acoustic problem for accurate hearing. Because air is much less dense than liquid, and far more compressible, without some skillful engineering most of the energy of the sound wave would simply be reflected back into the ear canal. A rough analogy would be throwing a rubber ball at the sidewalk. Most of the ball's energy is reflected in the ball's bounce back to the thrower. Very little is transmitted to the sidewalk.

For proper hearing, then, the body needs to amplify the signal between the eardrum and the cochlea. The best way to do this is with a lever system. Since the malleus is attached to the eardrum and the stapes to the cochlea, this leaves the middle bone, the incus, to serve as a lever. But not just any lever will do. Only a very specific configuration of that lever will properly translate the pressure waves in the air into corresponding pressure waves in the fluid.

The middle ear must provide a mechanical advantage to accurately bridge the different densities of air and fluid, and do so with minimal loss of loudness and tonality. Mechanical engineers call this *impedance transformation*, a tricky problem to overcome in even a simple system.

The ear's solution involves the precise shapes and configurations of all three bones of the middle ear. The malleus has a larger surface area than the stapes. Also, the two arms of the incus's lever have different lengths. Each provides mechanical advantage. Pressure waves hitting the large area of the eardrum are concentrated into the smaller area of the stapes so that the force of the vibrating stapes is nearly fifteen times greater than that of the eardrum. This makes it possible to hear even the faintest sounds.

These bones can only do their job effectively when surrounded by air. If they were immersed in fluid, the viscosity of the fluid would degrade their mechanical properties. This drives the need for an air supply to the middle ear.

The three bones of the middle ear, and the ways they're held in place by various tendons, act as a four-bar mechanism. The specific configuration in the ear is called a double-crank rocker. Engineers use four-bar mechanisms to fine-tune mechanical relationships in systems where exacting precision and sophistication are needed, as they are in the middle ear. To achieve the necessary mechanical advantage, the shapes of the parts and the positions of the several hinge points must be precisely tuned, with little room for error.

So, hearing relies on the precise configuration of these three tiny bones. Chapter 3 of our more in-depth book, *Your Designed Body*, explores how the very specific shapes of various bones are essential to their purposes. Nowhere do we see this more clearly than in the bones of the middle ear.

The Inner Ear

The inner ear is the most complicated part of the auditory system. It's so complicated that its exact workings are not yet fully understood. It consists of the bony labyrinth, a hollow cavity with a system of passages that includes the cochlea and the vestibular system. The cochlea is a helically shaped bony structure, something like a snail shell but with three parallel, fluid-filled chambers which spiral together for about two-and-a-half turns. The vestibular apparatus consists of the three semicircular canals, the utricle, and the saccule.

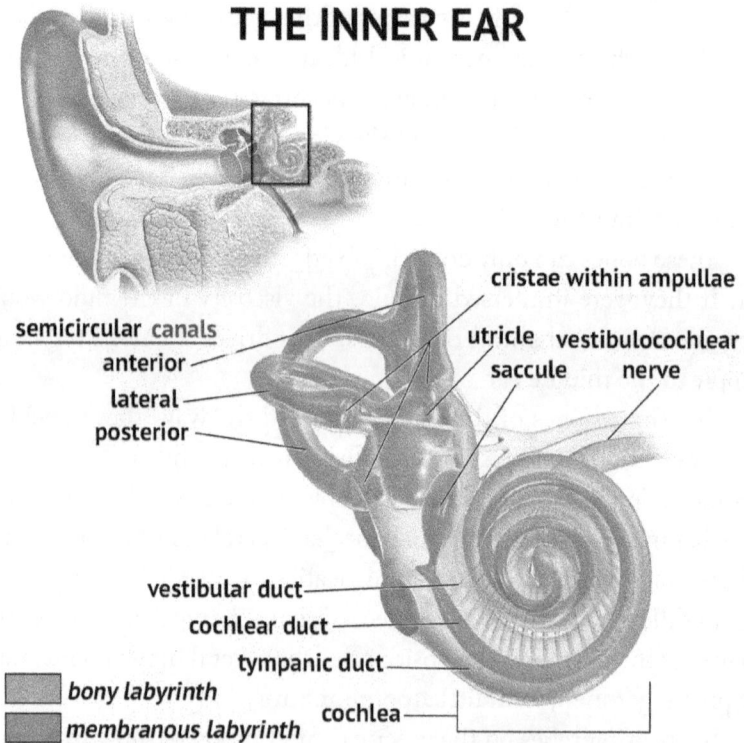

THE INNER EAR

cristae within ampullae

semicircular canals
anterior
lateral
posterior

utricle   vestibulocochlear
saccule   nerve

vestibular duct
cochlear duct
tympanic duct

bony labyrinth
membranous labyrinth   cochlea

Figure 6.3. The parts of the inner ear.

If the cochlea were uncoiled, it would be 3 to 3.5 centimeters in length. The three chambers that run its length are shown in cross-section in Figure 6.4. The upper and lower chambers are filled with a fluid called perilymph, chemically similar to the other extracellular fluids in the body, with high concentration of sodium ions (Na⁺) and low concentration of potassium ions (K⁺). In contrast, the middle chamber (the scala media), contains endolymph, a unique extracellular fluid with high K⁺ and relatively low Na⁺ concentrations, much like the fluids inside the cells. This is the only place in the body we're aware of where a fluid with this kind of chemical balance exists outside of individual cells. Though its purpose is not understood, it appears that this specific chemical composition is essential to hearing.

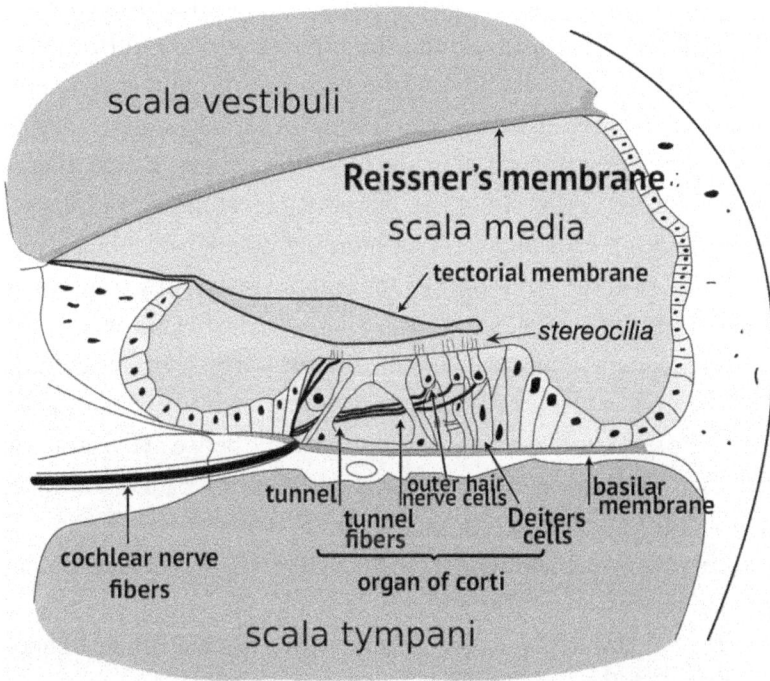

Figure 6.4. The three fluid-filled chambers of the cochlea.

Sound waves enter the cochlea's first chamber, called the scala vestibuli. The stapes acts like a piston to transmit sound waves to the fluid in this chamber through the membrane of the oval window. This sets up pressure waves in the fluid that correspond to the pressure waves that were in the air outside the ear. The pressure waves travel through this chamber from its base to the far end, called the apex.

Why is the cochlea of the inner ear coiled? Perhaps to better fit with the surrounding structures of the head. The human head crams a large array of systems into a remarkably compact space. The cochlea's coils may be the most compact design possible for a structure of this kind and complexity. In total, the human head presents a marvelous solution to a difficult packaging problem.

At the apex, the scala vestibuli connects through a small gap (the helicotrema) in the bony cochlear partition, into the third chamber, called the scala tympani. Here, the pressure waves return towards the base of the cochlea. At the base end of this chamber is a highly flexible membrane called the round window, which bulges in and out as the pressure waves move in the fluid. This allows pressure waves to be absorbed at the end of the chamber, preventing pressure waves from bouncing back and creating standing waves inside the cochlea.

Unlike air, the fluids in the inner ear are not readily compressed. So without the cochlea's pressure relief system, the ear would have little freedom of movement, and hearing would be severely degraded.

The middle chamber is where things get really interesting. This is called the scala media, since it sits in the middle between the other two chambers. A partition called the basilar membrane separates the middle chamber from the third chamber, the scala tympani. (See Figure 6.4.) As pressure waves pass through the third chamber, they cause the basilar membrane to bounce up and down, like a toy boat floating in a bathtub.

The basilar membrane is specially shaped to differentiate frequencies in sound. It tapers in width, from its widest (around 0.5 mm), near the apex at the helicotrema, to its narrowest (about 0.04 mm), at the oval window. This allows different wavelengths of sound (frequencies) to be maximized at different distances from the oval window.

The end closest to the base is tuned to higher frequencies, while the end closest to the apex is tuned for lower frequencies. In effect, the basilar membrane acts as an acoustic filter to separate the frequencies in sounds and direct them to different hair cells along the length of the cochlea.

The organ of Corti is attached to the basilar membrane. It has about 20,000 specialized "hair cells" which have thin bundles of fibers, called stereocilia, protruding from their top. A small bony "roof" called the tectorial membrane overhangs this area, such that when the basilar membrane moves up and down, the stereocilia are pushed back and forth at an angle against the tectorial membrane.

The hair cells differ in length by minuscule amounts and have different levels of resiliency to the fluid in the middle chamber. This means each hair cell is specially tuned to a particular frequency of vibration. When a sound wave's frequency matches the hair cell's natural frequency, that cell resonates with greater amplitude and induces the cell to release an electrical impulse into the auditory nerve, which sends it to the brain.

Thus, the basilar membrane's acoustic properties, combined with the fine tuning of the hair cells, produce our amazing ability to make fine distinctions in frequencies.

Battery-Powered Hearing

Hearing requires electrical signals to be generated in the inner ear and sent to the brain. For this the ear uses a remarkable mechanism that differs from anything else in the body.

In the inner ear, the organ of Corti's hair cells pump positive ions like sodium (Na^+) out of the cell, leaving a net negative internal charge of about -40 millivolts relative to the normal extracellular potential of the body. This is the hair cells' resting potential.

The stereocilia of the hair cells protrude into the middle chamber, which contains endolymph. Because endolymph is rich in potassium ions (K^+), it has a positive charge at around +80 millivolts. It's believed that this positive charge is generated by specialized cells that line the walls of the scala media. (Note that it takes yet another system to maintain the special charge needed in this fluid.)

Engineers have a special term for this kind of arrangement: It's called a battery, which in this case has its negative terminal inside the hair cell and its positive terminal outside, on the upper surface of the cell. When the stereocilia bend, the cell opens small ion channels, which allow K^+ ions to flow into the hair cells from the endolymph to equalize the electrical potential, in a process known as depolarization.

When the hair cell becomes depolarized, the voltage-rise in the cell opens specialized calcium (Ca^{2+}) channels, so calcium enters the cell, causing the hair cell to release a neurotransmitter.

Auditory nerve cells surround the bottom and sides of the hair cells. As the auditory nerve cell detects the hair cell's neurotransmitters, it "fires," opening its own ion channels at the synapse, which raises the voltage in the nerve fiber. A domino effect ensues, causing a depolarization wave to propagate along the nerve, all the way to the brain.

The inner ear's hair cells open potassium ion (K^+) channels, allowing K^+ ions to flood into the cell to depolarize it. This is unique and surprising, as every other nerve and muscle cell in the body (as far as is currently known) uses sodium (Na^+) ion channels for depolarization. We suspect this unique property will one day be shown to be essential to hearing. The fact that depolarization here uses such a different approach from that used in all other cells in the body poses yet another problem for evolutionary theory.

Protecting the Ears

Ears are sensitive instruments, so they're vulnerable to a range of threats. Fortunately, the ear possesses several defenses.

Perhaps the humblest of these defenses is ear wax, produced in the ear canal of the outer ear. Ear wax provides lubrication to keep the skin healthy while protecting the eardrum from dust, dirt, and invading microbes and insects.

But by far the greatest threat to hearing is damage to the tissues of the most sensitive parts of the inner ear. Remember, your tissues have physical limitations. Just as you can injure your elbow by hyperextending it, extremely loud sounds, such as a nearby explosion, can injure your eardrum, ossicles, and microscopic components of the cochlea by causing sudden and excessive movement.

Curiously, two skeletal muscles are located in the middle ear: the tensor tympani, attached to the malleus where it meets the ear drum, and the stapedius, attached to the stapes where it meets the oval window of the cochlea. (See Figure 6.2.) When the ear detects an extreme movement in the eardrum, a protective reflex instantly contracts these muscles, stiffening the sound transmission mechanisms of the ear, mechanically damping the transmission of sound through the middle ear to the inner ear. This protects the inner ear's most sensitive parts by reducing their relative movement.

Even though the physical reduction in amplitude is real, your brain seems to still perceive this as a noticeably louder sound. Surely this involves additional feedback systems that we don't yet understand.

Making Sense of Sounds

Auditory nerve cells react to signals from the hair cells in the inner ear and transmit them to the brain. There doesn't appear to be any difference based on frequency in the signals that different auditory nerves send to the brain, so as best we can tell the frequencies in sound are encoded based on the nerve cells involved. This is known as *place coding*.

Differences in amplitude are encoded in the rate the nerves fire. When the sound at a particular frequency is louder, the hair cells at that frequency fire more quickly, causing the associated nerves to fire more frequently. This is known as *rate coding*.

Between these two signaling mechanisms, both frequency and amplitude information are sent to the brain.

If place coding is how different frequencies are sent to the brain, distinguishing various frequencies must occur in the brain. How does the brain's hearing center know which nerve signals to associate with which frequencies? How does the hearing center know to interpret more frequent signals as louder volume? How are these connections assembled during fetal development? Where is this encoded in the DNA? (Or is it encoded somewhere else?)

While the body's solution to the challenge of getting sound-based signals to the brain is a good one, it does pose a problem. Around 20,000 hair cells from each ear can each send a signal to the brain at any given time. Each of these represents a different frequency, and the variations in firing rate represent different degrees of loudness. Given all this, the brain's auditory center has a lot of work to do in order to convert all this raw information into meaningful sounds. How your brain performs this feat is not yet fully understood. The bottom line is that assembling individual nerve impulses into coherent sounds is an enormously difficult problem.

Because the hair cells nearest the base of the cochlea respond to higher frequencies and the nerves closest to the apex respond to lower frequencies, the nerves mirror this arrangement, terminating in roughly linear order from front to back along the primary auditory cortex (the brain's hearing center). At least logically, this arrangement should make it a little easier for the brain to understand differences in frequencies. But of course, this complicates the assembly process during fetal development.

The Experience of Hearing

Finally, the mind needs to actually hear. To this end, it not only assembles all the individual nerve impulses into a real-time auditory whole, but also uses the fine nuances in the quality and direction of sounds to assemble a three-dimensional understanding of what's happening around the body. When this is coupled with visual signals, we're able to know the space around us with remarkable accuracy.

Further, our minds track and appreciate relationships in sounds over time. We know without looking whether a car is moving toward us or away from us. We understand spoken words and sentences by comparing many different frequencies and amplitudes over time. And we appreciate chord structures and transitions in music.

The rich bundle of capacities that is human hearing enables us to experience many nuances in volume, tonality, harmonics, timbre, cadence, direction, distance, movement, and more. Hearing gives us a richness of information about our world, and a richer way to experience the world around us. It's an amazing gift that enriches our lives.

Contemplating the gift of hearing should evoke wonder and gratitude. It should also prompt us to ask questions such as: Why do we perceive auditory information differently from visual information? Both are received as electrical signals by the brain, but they are translated into very different experiences. How do we know to do this? How do we do this?

Takeaways

The needs of hearing are simple to state, but incredibly difficult to achieve:

1. Capture acoustic pressure waves in the surrounding environment.

2. Amplify the most important range of the sounds, where importance is defined by the needs of the person, not the needs of the ear itself.

3. Convert the mechanical energy of those waves into electrical impulses.

4. Transmit those impulses to the brain without loss of information.

5. Generate a three-dimensional auditory "image" of the surroundings.

6. Animate this "image" to understand—and in many cases, enjoy—how the sounds are changing over time.

How many possible ways could hearing go wrong? Thousands. In almost all cases, a failure or error would severely degrade or eliminate the experience of hearing.

As with every system in the body, each part of the hearing system of systems must be in the proper position. The middle ear, for example, must transform waves in the air into waves in a fluid, and this is only managed when each bone has the correct shape, fits with its neighboring bones, and is held in exactly the right places and with the right pivot points. Only with extreme fine-tuning can mechanisms of this kind work properly.

Each part of the ear must be tuned for a specific purpose: The eardrum and oval window must have the right surface tension, the fluid in the middle chamber must have the right electrical properties, and so on.

The ear needs resilience mechanisms to protect against damage when sounds are too loud, else hearing would quickly degrade and you would lose both the information and the rich human experience hearing provides.

The shapes of the ear's many parts must have just the right acoustic properties. This includes the shape of the pinna, the diameter and length of the ear canal, the holes in the bones of the skull, the curling bony structure of the cochlea, the taper in the basal membrane, and many others.

Hearing requires specialized cells, and some of these cells, like the hair cells in the inner ear, must be precisely tuned to specific audio frequencies. Signaling between the ear and the brain must carry all the necessary information, and in a way that enables the brain to make proper sense of the sounds. And as with vision, the entire system must operate so quickly that you cannot perceive any time lag between the arrival of a sound and the mind's processing of it.

Hearing is an easy-to-understand example of how having the individual functions of the system is not enough. They must be strung together correctly in a continuous process from start to finish.

Human hearing requires solutions to many difficult engineering problems, and solves them in a strikingly elegant and efficient way, yielding acoustically accurate sounds, with an emphasis on the frequency range most central to human existence and human relationships. In short, our hearing system is an extremely clever (and near-optimal) set of solutions to a tough set of engineering problems.

7. Balance and Movement

You move because you are built for movement, because your
heartbeat and respiration are involuntary. Though you
feel the part of you that is always dying, the rest moves
forward, squeezed between destiny and choice.
—Cai Emmons[1]

THE PREVIOUS TWO CHAPTERS, ON VISION AND HEARING, WERE about senses that tell us about the world around us. But we also need to have a precise sense of where we are in our environment along with a fine-tuned sense of the spatial relationship of our limbs one to another moment to moment, and the ability to act and react based on that knowledge. Without all this, effective action would be impossible.

Your neuromuscular system solves this challenge using reflexes that help maintain posture, and active controls across hundreds of muscle groups to maintain balance. This wouldn't work without a lot of sensory information related to balance, and without ways to process the information quickly and effectively.

The required information comes from four main sources. These include visual cues from your eyes, the organs of balance in the inner ear, and pressure sensing in your feet. We delve into those in our longer book, *Your Designed Body*. Here we will jump to the fourth and most easily overlooked source—proprioceptors.

Proprioceptors

Your body can't control its actions if it doesn't know what its muscles and joints are doing. Proprioceptors are stretch receptors situated within your muscles, tendons, and joints. They detect changes in length and the rate of change using mechanically gated ion channels, and send this information to your brain. Some of these proprioceptors are responsible for initiating the automatic stretch reflexes that allow you to maintain your position and balance.

To maintain fine control over balance while moving, your body must employ its many individual sensors to continually feed positional information to the brain, and the brain must assemble these in real time into a holistic view of the body's position as it moves, adjusting future movements accordingly.

Here we see one of engineering's most useful tools: feedback loops. To control balance, the system's controller needs to monitor the effects of its commands and adjust them as needed. For example, when you reach to pick up an object, your mind estimates the force needed and sends appropriate commands to the muscles in your hands, arms, and core to maintain your balance. If the object is much lighter or heavier than expected, your brain detects this via feedback from those muscles and joints and quickly adjusts the signals it sends to the muscles.

Virtually all control systems in your body (actually, nearly all control systems everywhere) use one or more feedback mechanisms. Without them, your body would quickly spin out of control.

Coordination and Control

Building a body with the ability to maintain balance is a demanding engineering challenge, to be sure. But it's just table stakes in the world of movement. Another collection of hard problems must be solved to coordinate all the muscles and muscle groups needed to perform useful tasks.

Your musculoskeletal system consists of more than two hundred bones and over six hundred muscles. The muscles reach across dozens of joints. Skeletal muscles work by contraction and are controlled by motor nerves.

Since muscles work by contraction, a given muscle can move the eyeball or the bones of a joint in only one direction. To move them back requires another, complementary muscle. And to avoid a "tug of war," when one muscle contracts, its opposite must relax.

For complex bodily movements to work effectively, one other thing is needed: All the right signals must get to all the right muscles at exactly the right times.

Each skeletal muscle consists of numerous muscle fibers. When stimulated by a motor neuron, the muscle fibers contract, and the angles between the bones that the muscle is attached to will be changed. The group of muscle fibers controlled by a given motor neuron is called a motor unit.

Different muscles have different numbers of muscle fibers in each of their motor units, according to their purpose. For the coarse strong movements of the back, legs, and arms, a single motor unit will have from hundreds to several thousand muscle fibers. For the fine and precise movements of the eyes and fingers, a given motor unit might have as few as five to ten muscle fibers.

More refined muscle control takes more refined feedback, so more positional sensors (muscle spindles) are embedded in those muscles that need finer and more precise action, while fewer are present in muscles that need only coarser and less precise action.

But there's still more to motor control. The movements of the body need the simultaneous coordination and control of multiple muscle groups—a choreography of movement.

Touch your nose with your left index finger. What's the actual complexity of this "simple" action? How many body parts move, which muscles do what kinds of activities, and what is the timing and relative strength required from each muscle involved? Despite its great complexity, your mind coordinates the many needed signals without conscious controls.

For this to work, the brain must precisely coordinate different signals to many different muscles at the same time. If this kind of coordination were missing and the many muscles involved were not given coordinated signals regarding contracting and relaxing, the body would be immobilized. If even one of the components of these movement systems were missing, the body would be significantly hobbled.

Imagine being able to extend your arm (push) but not bring it back in (pull). You need both. Again we come back to the fundamental question: If this richly complex system of systems needs all these elements in place for even basic function, how could it possibly have evolved one small step at a time?

8. A THEORY OF BILLIONS
OF INNOVATIVE ACCIDENTS

*I look with confidence to the future, to young
and rising naturalists, who will be able to view both
sides of the question with impartiality.*
—CHARLES DARWIN[1]

SOLVING A GIVEN PROBLEM IS RARELY SIMPLE. IT'S COMMON FOR a solution to one problem to present new problems that must also be solved. And each of those solutions likely presents still more problems.

Doctors constantly watch for this. A treatment regimen for a given condition often causes other issues that must also be addressed—called iatrogenic (doctor-induced) complications. For engineers this challenge is also business-as-usual. Most problems have many possible solutions, and most solutions present new problems, which then must be solved, generating still more problems, and so on, until adequate solutions have been engineered across the whole. Only when all the solutions are brought together, and the rabbit-like proliferation of problems-from-solutions is finally corralled, does the resulting system work.

We see the problem of cascading problems all over the place in the human body. The oxygen supply required for cellular respiration illustrates this well. Figures 8.1a and 8.1b illustrate how these problems pile up, and how each must be solved in turn to solve the initial problem of cellular respiration in the human body.

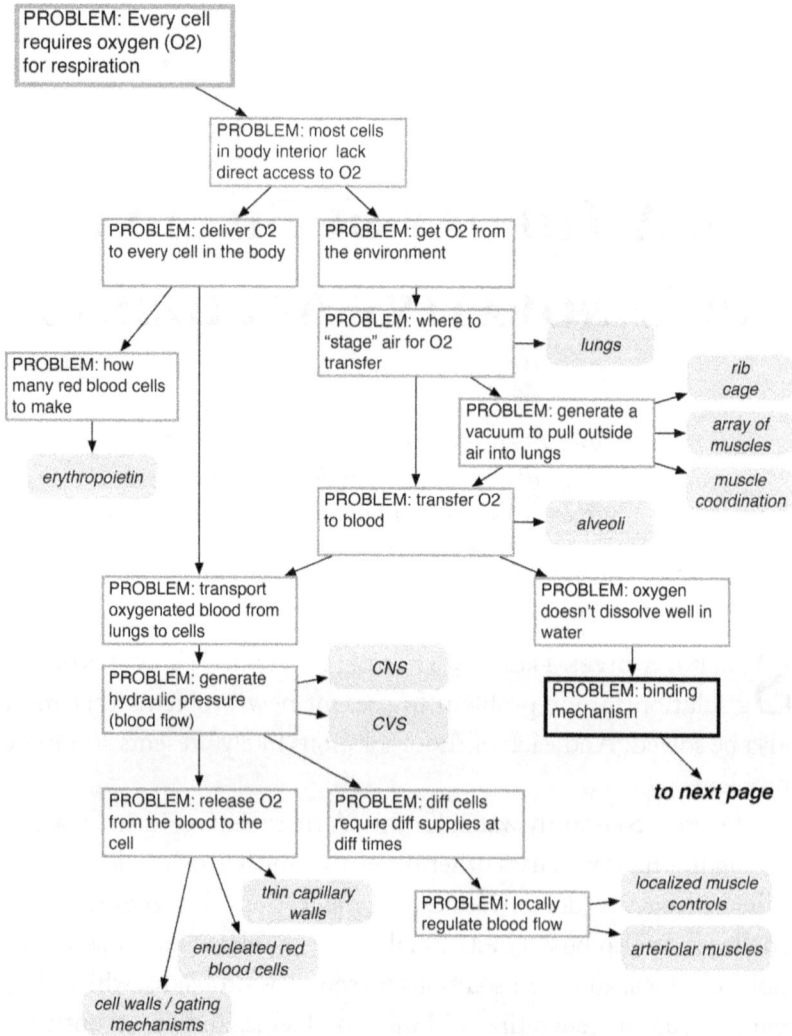

Figure 8.1a (above) and Figure 8.1b (right facing page). A problem cascade for cellular respiration in the human body. Problems are listed in boxes, and solutions in ovals. This illustration greatly simplifies, and therefore understates, the complexity of the problem cascade. So, for example, several of the boxes are "stubbed out" (e.g., those labeled "GIS" and "CVS") and are best viewed as markers to another distinct but similar problem cascade, which must also be solved in order to solve the initial problem.

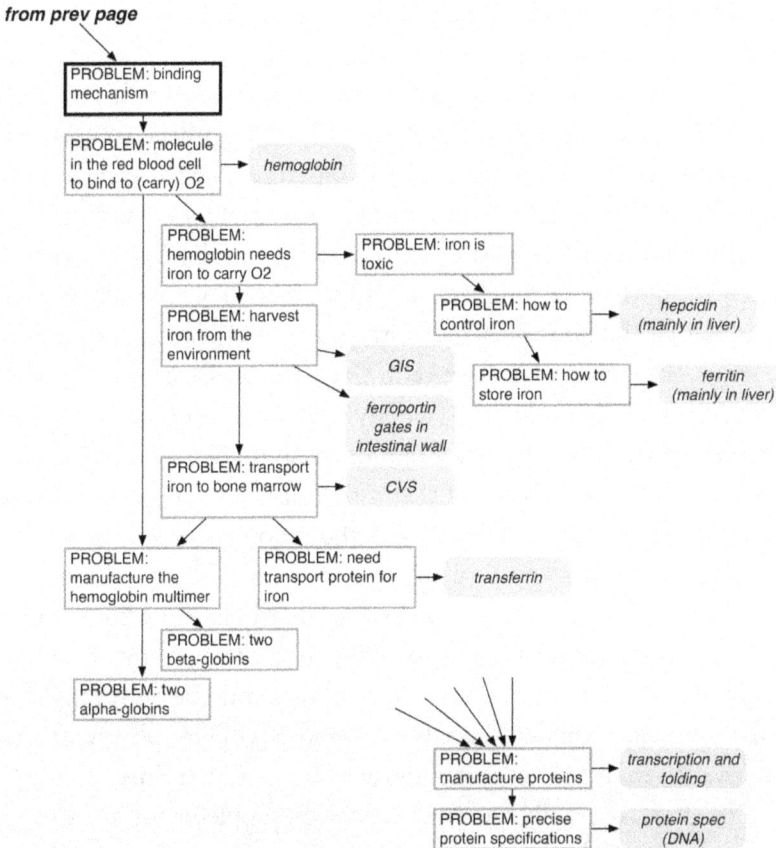

Figure 8.1b. Problem cascade for cellular respiration, continued.

Remember, solving the cellular respiration challenge is not just a wouldn't-it-be-nice engineering challenge to solve. Every cell must get oxygen or die. Cells that can't live don't do the body any good, and if enough fail, the body also fails. It's essential to get oxygen to every one of those thirty trillion cells, but in a large body very few cells have direct access to oxygen from the atmosphere. This is a demanding problem. Even just this high-level view of the problem cascade identifies more than two dozen problems, each requiring a different complex solution composed of many specialized parts acting in a coordinated way.

Imagine you are asked to build a human body. You're given all the carbon, oxygen, water, nitrogen, metals, and other chemicals you'll need. How would you solve this engineering problem? If you don't solve the cellular respiration problem and all the others, you just end up with a pile of ingredients, not a living, functioning creature.

Respiration is not an isolated case. We could start with almost any problem the body must solve to be alive, and chase the cascading problems until we've touched most of the body's other systems. This tells us that the body is loaded with interdependencies. Most of the systems only work when they're in the context of the other systems they rely on. What follows? If a body must solve a couple of hundred problems to be alive (the actual number is far higher), and each bodily system solves one or more of them, then it takes the whole body to solve all the problems. Were any of those do-or-die problems to go unsolved, the body wouldn't survive.

This is important. When we look at the whole of the body, which of the systems, or subsystems, or sub-subsystems, or any of the underlying information, could be taken away before the body fails? If it can't function without hundreds and hundreds of key subsystems and parts, how could it have come to exist a little bit at a time?

In short, the body has chicken-or-egg problems top to bottom. Indeed, most of the problems the body must solve are chicken-or-egg problems. And this is not an arbitrary difficulty based on some quirky approach to engineering the human body. In every case it's a natural outcome of the laws of chemistry and physics. The body must solve these problems and solve them well if it is to be alive and thrive.

Could material causes alone have solved such problems, given enough time in the history of life, as proponents of modern evolutionary theory assert? In answering this question we must confront the reality that nearly every important system or subsystem in the human body is irreducibly complex and works coherently to achieve ingenious, fine-tuned solutions to the hard problems of life. By "irreducibly complex" we mean that for the system to function, the suite of solutions to the subproblems must be complete. Put another way, some minimum number of parts or controls are required for the system to work. Take any one of these away and, like a simple mousetrap missing one of its five parts, it doesn't work at all.

The bottom line is that pretty much everything essential or meaningful in the body is irreducibly complex. So how could a gradual evolutionary process build such intricate systems, one small mutation at a time? Evolution can't wait for future function once all the essential parts have been assembled. It's now or never. Do or die.

The challenge posed in Figures 8.1a and 8.1b is the rule rather than the exception. The complexity of the system described in that greatly simplified two-page illustration is only hinted at there, and even if we could somehow have crammed a description of all the complexity of that system onto those two pages, it would still have represented only the tip of the iceberg. The body is brimming with complex interdependencies across its thousands of subsystems and between the layers of its design hierarchy. If we're ever to understand the origin of the human body (or of pretty much any other living thing), we'll need a cause, or combination of causes, capable of generating these types of systems and solutions.

Darwinism and Its Tools

In 1859, Charles Darwin published *On the Origin of Species*, his seminal work on evolution in the history of life. He updated this with five more editions over the next dozen years, responding to various critics and refining the theory. Others have updated and modified the theory since, notably with the development of population genetics in the 1930s, followed by major discoveries about DNA in the 1950s and subsequent decades. The modern form of the theory is known somewhat fluidly as neo-Darwinism, a term that overlaps with an even more plastic term, modern evolutionary theory, which includes some variations on Darwinism known as the extended evolutionary synthesis. There are many variations and nuances, but they all share the same basic principles and the same causal limitations.

Darwin's theory states that the biosphere's great diversity springs entirely from natural selection acting on random variations. As an organism generates offspring, small changes can occur, yielding the process of change known as descent with modification. Some of these changes may confer an increased survival or reproductive advantage to the offspring. Over time, traits that offer an advantage are more likely to become fixed in a population, and individuals lacking that advantage will likely die out of the population. Darwin termed this natural selection, where material causal forces work like farmers who select which animals or plants to breed to produce desired effects in the offspring. However, in Darwin's world, unlike the world of animal breeders, selection of random traits occurs without planning or foresight toward any distant goal.

While there are additional details we won't go into here, Darwin's theory primarily involves the interplay of four causal factors, all of them strictly material causes, devoid of purpose and planning:

- *Variation*: A means by which differences occur in an organism. Darwin himself did not propose a mechanism for variation, but since the discovery of DNA, these changes are known as mutations and are mostly attributed to random errors in processing or copying the information in DNA.

- *Heritability*: A way for an organism to pass its traits, including its variations, to its offspring. This leads to the notion of common descent, wherein all living organisms are said to have descended from a single common ancestor, known as the last universal common ancestor (LUCA). The branching caused by descent with variation is often drawn as a tree of life, showing all living things descending from the hypothetical LUCA. (Contrary to the expectations of modern evolutionary theory, the data do not fit a clean and unambiguous tree-of-life pattern. For this reason, among others, no one has been able to reverse engineer the traits that the LUCA might have had, much less confirm that it ever existed.)

- *Natural Selection*: In the struggle for survival, certain organisms, given their variations, are able to out-compete others in a given environment. Individuals with better survivability tend to survive, and the others tend not to. This is known more colloquially as survival of the fittest.

- *Time, Lots of Time*: The theory depends on eons of time. Gradual, intention-free, trial-and-error evolutionary progress just doesn't happen overnight.

These four factors constitute the major drivers of an evolutionary theory of life *gradually innovating and diversifying by accident*. As a strictly materialist causal theory, Darwinism (and its modern variant, neo-Darwinism) has no place for intention or foresight: no room for goals (beyond an organism's impulse to survive and reproduce), no source for guidance in whether or how to assemble a system, and no means to discern whether a partial system will be functional in the future, in the event it should ever be completed. No one is available to intend any particular outcome, evaluate partial progress or, guided by foresight, steer things in a promising direction. The only possible way forward is by trial and error.

Some Attractions of Darwinism

Though a few materialistic theories of biological origins preceded Darwin's, his was the first to propose a specific mechanism for producing change over time in a population. As there are countless examples of change over time, and no other seemingly viable natural mechanisms were proposed, his theory gained traction in the scientific community and eventually spread to the general population.

The theory has an attractive simplicity and elegance. It's easy to understand and on its surface seems plausible. It addresses changes in the fossil record, including the observation that life-forms have become more complex over time.

Further, for those who desire the materialist worldview to be true, Darwinism offers three attractive perks. First, for many people, one of science's primary goals is to replace superstition with natural explanations. For them, any designer-engineer, especially a transcendent one, is a quaint superstition that should be displaced by materialist science. Darwinism restricts itself to natural explanations. Second, Darwinism is an enabler for scientism, the belief that science is the only path to true knowledge. This view is desirable to many people, particularly scientists. Third, the theory is immensely attractive to those who would prefer that no God exist. As Richard Dawkins put it, "Although atheism might have been logically tenable before Darwin, Darwin made it possible to be an intellectually fulfilled atheist."[2] The desire for facts that align with a preconceived worldview can be a powerful motivator.

Darwin's Difficulties

The factors above help explain why Darwinism persists. But the theory also has deep-rooted issues and limitations. Darwin discussed some of these in the *Origin*, in a section titled "Difficulties on Theory." Subsequent generations of biologists, including ones sympathetic to the project, have also pointed out problems with the theory. Here is a list of some of the biggest ones.

Innovation Goes Missing

The first and most glaring difficulty with Darwin's theory—both its original and updated forms—is that, as Gerd B. Müller and Stuart Newman put it in an MIT Press book, "It completely avoids the origination of phenotypic traits and of organismal form. In other words, neo-Darwinism has no theory of the generative."[3] Müller, a leader in the extended-evolutionary-synthesis movement, says that "selection has no innovative capacity" and "the generative and the ordering aspects of morphological evolution are thus absent from evolutionary theory."[4]

In other words, nothing in the theory can generate any non-trivial innovation. There's simply no combination of Darwinian causal forces or factors that can overcome any causal hurdle of more than trivial complexity. Even combined, and with billions of years at their disposal, Darwinism's causal forces are much more likely to destroy potentially helpful changes long before enough changes can accumulate to add a useful new function.

Like the Mail, the Complexity Just Keeps Coming!

In the '90s sitcom *Seinfeld*, the mailman Newman (a fictional character, not to be confused with Stuart Newman, the biologist just cited!) explains why, from time to time, a mailman flips out and goes *postal*. "Because the mail never stops," he says. "It just keeps coming and coming and coming!"

Like the mail, new discoveries regarding the true complexities of living systems just keep coming! And this has steadily eroded the plausibility of Darwin's causal explanations. For example, in Darwin's day no one knew much about the workings inside a cell. His go-to expert on cellular biology, G. H. Lewes, described the cell's internal contents as a "microscopic lump of jelly-like substance" with "no trace of organization."[5] Darwin accepted this view and in a private letter proposed that the first living organism arose by chance in some "warm little pond."[6] Lewes's understanding of the cell couldn't be further from the truth. The cell is a fantastically complex factory with millions of specialized parts, precisely organized and perfectly orchestrated.

And that's just a cell. The human body has layers upon layers of coherent interdependent systems, and with each new discovery, we uncover more functions, more information, more coherent systems, more interdependencies, more fine tuning.

Each new answer raises still more questions for evolutionary theory, but Darwin's cupboard is bare. The theory's causal tools are spent, unable to overcome the sheer force of the evidence building against it.

Bounded on All Sides

In the *Origin*, Darwin offered no specific mechanism for variation. He assumed that living organisms have unbounded plasticity—that they can gradually change from anything into anything.

It's true that information storage in DNA is practically unbounded. As a four-value digital code, it can hold any digital information. Because of this and its extreme compactness, computer scientists are researching ways to use DNA as a computer storage system. But it's not true that just any random combination of information in DNA will code for a living system. The hard problems that life must solve act as bounding constraints on life. Life can only exist in a tiny subset of the total possible arrangements of DNA or RNA or other information codes. Rather than possessing unbounded plasticity, then, life is effectively bounded on all sides. And this makes it extremely problematic for one organism to change into a completely different organism, regardless of the time allowed. There's simply no way to be alive in the in-between stages.

A Predictions Scorecard

Darwin's theory has also faltered in its predictions. For example, Darwin predicted large numbers of intermediate forms, even though in his day the fossil record suggested that most new body plans appeared suddenly, usually with no observable precursors. He placed his hope in future fossil discoveries to bring to light the vast number of intermediate forms that his gradualistic theory predicts, and thereby validate it. Instead, more than a century and a half of additional fossil discoveries has only confirmed the pattern that challenged Darwin's theory in the first place.

Counterintuitive

Darwin's theory is and has always been counterintuitive. Many people have an innate sense that there's more to a living creature than Darwinism alone can explain. It's hard to imagine in detail how any series of undirected accidents, in the absence of intention, could build anything like the human body, or even a molecular machine such as an ATP synthase. Obviously, common sense is not the final arbiter of truth, but it suggests that there may be a fundamental weakness in the theory and that, at the very least, it should not be embraced dogmatically.

Selecting for Death

Darwin showed a certain rhetorical genius in coining the term "natural selection"—implying agency, even while disclaiming it. The term was intended to explicitly place nature in the role of an animal breeder (an intentional actor). But the breeder's "artificial selection" is intentional.

In fact, there seems to be a limit to how far a creature's genes can bend before they break. Some species, like dogs, have enormous plasticity in their genes, but no dog breeder has so far made a distinct new species, much less an entirely new body plan. In fact, there's evidence that many of the variations in modern dogs are due to genetic degradations from the wild type they're derived from.

In evolutionary terms, nature mainly does one thing—it constrains life. (To be sure, the laws and constants of nature are fine-tuned to make life possible, but they can do nothing to make life in the first place.) The environment, rather than shaping living things, simply provides the set of problems that life must solve. It can do nothing to help organisms solve those problems. Instead, nature drives all systems toward equilibrium. Equilibrium sounds peaceful enough, but as we've seen, equilibrium with the environment means death—not just for the human body, but for all living things.

So, to the extent that nature is able to select for anything, it selects everything—for death. Only the internal capabilities of organisms can prevent that. And even then, only for a while.

Hard Problems

Given the hard problems we've presented in these pages, it's clear that Darwin's causal factors have two other problems, which together seem insurmountable.

First, the two hardest problems life must solve—maintaining a separate equilibrium (being alive) and reproduction (making a living copy)—must be solved *before* Darwinian causal factors can do anything. As we've shown, these are never trivial problems. Not even for the simplest bacteria.

Second, on the whole, Darwin's causal factors are not the solutions to life's problems; they are themselves problems that life must solve. Life must solve the problems that nature presents. It must be able to make a living copy of itself. It must overcome the degradations of random mutations. And it must thrive over long periods of time, even when entropy beckons.

Gradualism Revisited

Darwinism's commitment to gradualism is dictated by Darwin's four causal factors. If his unguided joint mechanism is going to create anything, it has to be through a long series of small steps. But we've shown that systems and subsystems in the body profoundly challenge gradualism. In particular, no system with non-trivial coherence can be built gradually if it has to be alive at each step along the way, because it won't live long enough to reproduce.

For example, how would a creature control its salt content while it's waiting for a complete salt control system to gradually evolve and finally come online? Out-of-control salt is a killer. How many generations would be required to persist without a functioning salt control system?

Logically, the only exception is where a protective system is in place to keep partial solutions from being destroyed while the whole waits to be completed. This by itself is a tall order. And more to the point, it smacks of foresight, planning, and clever engineering—reviving the problem Darwinism was supposed to solve in the first place.

Coherence is a qualitative problem and holds true regardless of the amount of time that may or may not be available. While gradual causation may be able to make small-scale tweaks to an existing system, it can never overcome non-trivial coherence to make a new coherent system.

As a causal process for the creation of new biological systems, then, Darwinian gradualism fails.

Growing Discontent in the Ranks

Issues like these are thorny and increasingly hard to ignore, leading a growing number of evolutionary biologists to seek out add-ons or extensions to neo-Darwinism and even to openly express their discontent with it.

Several variations have sprung up from these efforts to rescue modern evolutionary theory. Could one of them overcome neo-Darwinism's limitations? A detailed exploration of these variations takes us beyond the scope of this book. But consider one popular variation, called neutral evolution, which relies less on natural selection. Instead, it proposes random, non-selectable allele variation and genetic drift as the main drivers of change, at least at the molecular level. In essence, this approach posits that a lot of variations (alleles) in a given gene are neutral with respect to natural selection, so the sequence of the gene can take on different states and randomly drift in a population. These changes do not require a "selective advantage."

But this elaboration on Darwin's theory only addresses the molecular level, and downplays the only natural force available to Darwinism to preserve useful features: natural selection. In doing so, it leaves no reason why useful traits should evolve other than dumb luck. Dumb luck is never a good mechanism, especially when you're trying to engineer machines and systems more complicated than anything ever built by a human.

Indeed, by largely dispensing with natural selection, the neutral model dismisses the very thing about Darwinism that raised it above all the previous speculations about unguided evolution and provided a way to filter random variations and nudge a population on an upward trajectory. Evolutionists, it seems, can't seem to live with natural selection, and they can't live without it.

They face a pick-your-poison dilemma. Without natural selection, random variations wander aimlessly in astronomically big sequence space. With natural selection, organisms cannot take a vacation from functionality while a new organ or system gradually evolves over many generations. They must maintain function at every step, a demand that proves a dealbreaker for blindly assembling the coherent and interdependent systems of systems required of even relatively simple life-forms. Attempts to forge some sort of hybrid model, with neutral evolution doing some of the work, and natural selection swooping in as needed, have thus far failed.

The details of neutral theory are complicated, and the evolutionary community is still debating its capabilities, but there's nothing here that can overcome the causal hurdles described in these pages, or for similar causal hurdles in any other type of organism for that matter.[7]

This is typical of the variations on neo-Darwinism. They tend to focus on modest variations on the broader themes introduced by Darwin, and none have any tools with greater capabilities or lesser limitations than what we covered above. And all remain limited to purely material causes.

The situation is becoming increasingly awkward as the evidence continues to skew away from Darwinian materialism. For those with an allegiance to both the evidence and that framework, it's like the guy who, after untying his boat, finds himself with one foot on the dock and one foot in the boat. As the gap grows, the situation becomes increasingly uncomfortable, and the need to choose, increasingly hard to ignore.

In the search of a new theoretical foundation, in 2016 the Royal Society convened a scientific meeting in London to explore neo-Darwinism's deficiencies and explore new theoretical possibilities—what some call a "third way,"[8] one that sticks to purely material causes but that isn't restricted to the neo-Darwinian causal mechanisms.

Gerd Müller, emeritus biologist at the University of Vienna, opened the conference by asserting that the modern synthesis "does not explain" what he terms "complex levels of evolution," including "the origin of... body plans" or "complex behaviors, complex physiology" and "development." According to Müller, the modern theory of evolution "is focused on characters that exist already and their variation and maintenance across populations, but not on how they originate." As we saw earlier in this chapter, in his view the standard model is not even "designed for addressing" questions like the origin of novelty.[9]

In a similar vein, University of Chicago microbiologist James Shapiro and his colleagues comment at the Third Way website that neo-Darwinism "ignores much contemporary molecular evidence and invokes a set of unsupported assumptions about the accidental nature of hereditary variation." They continue:

> Neo-Darwinism ignores important rapid evolutionary processes such as symbiogenesis, horizontal DNA transfer, action of

mobile DNA and epigenetic modifications. Moreover, some Neo-Darwinists have elevated Natural Selection into a unique creative force that solves all the difficult evolutionary problems without a real empirical basis. Many scientists today see the need for a deeper and more complete exploration of all aspects of the evolutionary process.[10]

At its end, the Royal Society conference had laid out many detailed challenges to neo-Darwinism but came up empty in the search for a new theoretical foundation.

A Few Thoughts

Not only did Darwin frankly discuss "Difficulties on Theory," but he also offered a way to falsify his theory: "If it could be demonstrated that any complex organ existed, which could not possibly have been formed by numerous, successive, slight modifications, my theory would absolutely break down. But I can find out no such case."[11]

A lot has changed since Darwin's day. Where he could see no such case, in this book, we've explored many examples in the human body of systems that cannot reasonably be attributed to any gradual evolutionary process, in any time frame, even if they were designed. In fact, given the kinds of causal hurdles in the human body and in all other living things we know of, we can go further and say that no one has yet identified *any* organ or system of significance, in all of life, that could have been created gradually, one small step at a time.

In the end, Darwin failed at the main thing he set out to accomplish in his 1859 work—to explain the origin of species. Darwinism, together with every other theory that relies on intention-free gradualism,[12] has failed—by Darwin's own explicit test. Or as Michael Denton put it, only slightly more gently, evolution is "a theory in crisis."[13]

But when we let the evidence determine what is and isn't acceptable rather than being guided by a question-begging methodological rule, we find a much more interesting—and as we will argue, fruitful—way to think about these things.

9. But What About All Those Botched Designs?

If they can get you asking the wrong questions,
they don't have to worry about answers.
—Thomas Pynchon[1]

SOME LOOK AT THE HUMAN BODY AND, RATHER THAN BEING amazed by its extraordinary engineering, focus instead on what they consider indications of botched design. Such arguments are typically marked by drive-by criticisms of various aspects of the human body—e.g., Kenneth Miller assuring us of "the many imperfections of the human backbone,"[2] or Nathan Lents insisting on the pointlessness of most of the ankle's seven bones,[3] or Richard Dawkins informing us that the retinal structure of the mammalian eye "would offend any tidy-minded engineer."[4]

These and similar arguments have a pedigree stretching back at least as far as Charles Darwin himself. In one of his letters he suggested that one reason he preferred his theory to a design framework was his conviction that the living world is riddled with misery and bad design. Writing to his friend Joseph Hooker in 1856, he commented, "What a book a Devil's chaplain might write on the clumsy, wasteful, blundering low & horridly cruel works of nature!"[5]

Cruelties in nature have long puzzled many who are committed to the idea of a good Creator closely involved in his creation, but these cruelties struck Darwin as easily explained on the grounds of mindless evolution by random variation and natural selection, a process that did not involve God.

The cases of apparent poor design that Darwin and subsequent Darwinists have flagged are certainly worth attending to. The problem isn't that they attend to these matters; the problem is that they attend to them superficially.

The Pathetic Design of the Bad-Design-So-No-Designer Arguments

Specific "bad-design" arguments in biology have a habit of collapsing under scrutiny, but even before this, one can recognize a more basic weakness in these arguments. Even if the human body did possess certain features that were poorly made, those features might still provide clear evidence that a designer-engineer had fashioned them. The Yugo automobile was infamous for its shoddy quality. But even a cursory look at it, even from someone who had never seen an automobile, would reveal it to be the product of intentional design. Neither the parts nor the whole worked very well, but shaping, selecting, and coherently arranging the parts was clearly the work of design.

Logically, even a bona fide example of a badly designed system in the human body couldn't erase all the examples of masterfully engineered systems reviewed in these pages. Nor could it erase the causal challenges these systems pose to any intention-free cause. Those challenges demand real answers.

The point of bad-design arguments is to call into question the motives or competence of the designer, and ultimately, to conclude that the system was therefore not designed. This has been called "presumptive theology" because it asserts (typically without support) what God would or would not do. It's a theological argument (though usually cloaked beneath layers of scientific jargon), and not a very good one at that.[6]

If faced with an argument like this, ask the critic to provide a workable improvement on the design in question, one that achieves all the capabilities of the original and hews to real-world constraints, solving all requisite engineering problems across an organism's complete lifecycle, including its development in the womb. This is a lot harder than it looks. Frankly, if anyone can manage it, a likely fortune awaits in industry. Per the above example, if you can design a better ankle than the one in the human body, the robotics industry wants to hear from you. The market potential is enormous for improvements, especially for medical implants and in building better robotic systems.

Among those putting forward a "poor-design-so-no-designer" argument, there is generally little effort given to make a serious case. Instead, the argument usually consists of storytelling, often as a form of entertainment for the neo-Darwinian faithful.[7]

With even a little creativity, one could apply this style of argument to any design, by any designer-engineer, anywhere. Take an example from my early years. The town where I (Steve) grew up was a lumber town. Chainsaws are remarkably helpful in the process of turning trees into lumber, but they're also noisy, smelly, and extremely dangerous. It would be easy to write about how poorly designed chainsaws are, with toxic exhaust fumes (making humans sick), high noise levels (damaging human hearing), and an open blade (removing a human limb as fast as it can remove a tree limb). But how could the benefits of a chainsaw be achieved with a different system? The power-to-weight ratio of its noisy two-stroke engine makes it portable to carry into rugged terrain, yet with enough power to do its job. Putting a guard over the blade would make it safer but render it unable to achieve its purpose. Adding a heavy-duty muffler or an aggressive smoke filter would compromise power and make it heavier.

Given that every real-world system is open to blinkered criticisms that ignore trade-offs, are we to conclude that nothing is designed? Of course not.

Here are a few examples of supposed poor design in the human body:

- The "backward wiring" of the vertebrate retina, leading to a blind spot.
- Various parts of the musculoskeletal system, including the ankle/foot, spine, knee, shoulder, elbow, and wrist/hand.
- The exposed testicles of human males.

- The narrow pelvis and birth canal in human women.
- The structure of the sinuses.
- The circuitous path of the left recurrent laryngeal nerve.
- The inability of humans to manufacture their own vitamin C.

Some of these arguments are approaching urban legend status. We briefly addressed several of these in our longer work, *Your Designed Body*. We'll touch on just two such arguments below—the supposed inadequacies of the human wrist and the choking hazard in the human pharynx.

Much more can and has been written exploring the failings of these "botched-design" arguments.[8] Such arguments are unlikely to impress actual designer-engineers of complex systems, who are accustomed to juggling multiple design goals, threading the needle between unforgiving parameters, and choosing among difficult design trade-offs.

Discordance

A closely related but more serious counterargument to our designer-engineer thesis is called discordance. We find this in Chapters 10–14 of Darwin's *Origin of Species*—roughly the final third of his original treatise. There he developed a special form of his case for evolution, what Stephen Jay Gould called "one long list of examples for inferring history from the oddities and imperfections of modern objects."[9]

As Darwin explained, the key clues were in "organs or parts in this strange condition, bearing the stamp of inutility."[10] The logic of the discordance arguments goes something like this: If an organism has quirks, oddities, or imperfections, or if some part of it is poorly coordinated with its current circumstances (environment), but these features align with readily inferable previous circumstances, then it's reasonable to conclude that these features are holdovers (vestiges) from those earlier circumstances. Or, stated more simply, one can infer that an organism's external environment changed so quickly that it hasn't yet been able to evolve to fit the new environment.

Thus, discordance requires two steps: (1) find a feature or function that doesn't fit well in an organism's current circumstance, and (2) show that this feature did fit well in a clearly inferable past circumstance. So-called "vestigial organs," like the human appendix, are commonly cited. But these arguments have fared badly over time. Virtually all the "vestigial organs" have been shown to have an important present purpose.[11] "Junk DNA" is a similar case. It has been widely cited as evidence of evolution at its most inept, but researchers are discovering crucial roles for the stuff.[12] As of this writing, the idea of junk DNA has been so thoroughly debunked that there's now little talk of it in the research community.

Another example in vogue involves "diseases of civilization" because the diseases seem to result from rapid changes in Western society, especially the Western diet, for which the human gastrointestinal system has not had time to properly adapt. (Presumably, in a couple million years our systems will evolve and it'll be OK to eat donuts.)

Such arguments are even worse than they may appear at first glance. Think about it. They actually are arguments that something has *not* evolved to fit a new environment. But to succeed, evolutionary theory needs to produce evidence of evolution happening, and of its inventing things far more novel than, say, an improved capacity for metabolizing donuts.

Discordance is also invoked to explain *homologies*—where similar structures are used in very different ways in different organisms. For example, the bone structure of the human wrist is structurally similar (homologous) to the wings of bats and birds, the front flippers of whales and dolphins, and the forelegs of dogs and crocodiles. In these cases, the bones and their relative positions are roughly the same, but their shapes, relative sizes, and optimal uses are quite different.

Darwinists claim that this must be the result of common descent. But that's not the only available explanation. A possibly better explanation is common design. The best designers reuse facets of their designs—what engineers call design patterns.

Don't Sleep on the Wrist

In one instance of an almost comically lazy discordance argument, Nathan Lents insists that the human wrist "is way more complicated than it needs to be" and that its bones are "like a pile of rocks—which is how useful they are to anyone."[13] In other words, the wrist bones are an evolutionary accident that the human body inherited from some distant ancestor, then gradually evolved into a working hand, but these specific bones no longer have any use to us.

But are these bones useless? The only way to answer with rigor is to perform a mechanical analysis of the range of movement and the stresses imposed by the various activities the hands and wrists are used for. Is it possible that every bone in the wrist serves an important purpose? It turns out that the answer is "yes." Kinesiologists have long understood the basics of how the human wrist works, and not surprisingly, every bone in the wrist serves a useful function. Notably (and amazingly, from the engineering point of view), the wrist contains a double hinge, in which two different joints share the same axis of rotation, allowing a wide range of motion, yet with incredible strength. Those "useless" bones form these hinges.

In any serious evaluation, the many structures and capabilities of the human wrist, including the many ways its joints are hinged to move in three dimensions, provide a lesson in the highest possible design prowess.[14] Rather than a useless "pile of rocks," what we find is an amazingly fine-tuned system of structure, flexibility, strength, and dexterity. Instead of discordance, we see superb coherence, a system precisely tuned to exactly match the human body and make it even more capable than it would otherwise be.

Additional research into the mechanical capabilities and functions of the wrist is ongoing. We are confident it will uncover still more evidence of ingenious engineering.

The Supposed Bad Design of the Human Pharynx

Or consider the claim that the human pharynx (throat) is poorly engineered. The pharynx is the common entry for both the respiratory and gastrointestinal tracts. Whatever is ingested can potentially go down the airway and cause obstruction, which can result in death by choking.

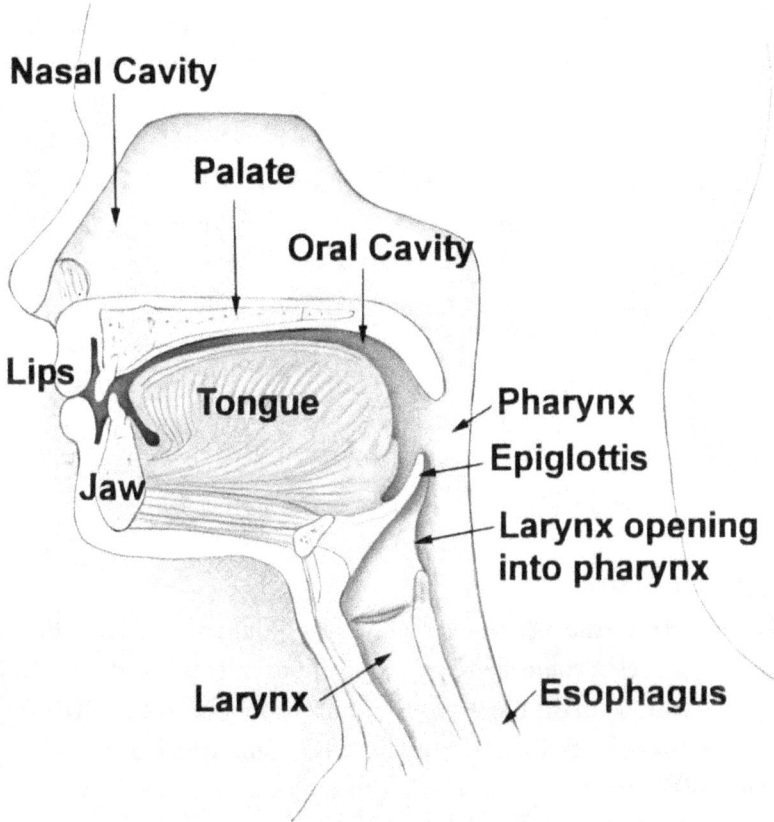

Figure 9.1. The anatomical relationship of the nose, mouth, and throat, showing that the pharynx is a shared pathway for air to enter the respiratory system and for liquids and solids to enter the gastrointestinal system.

Some insist that the pharynx is therefore miserably designed, something no wise designer would engineer, but that evolution, with its trial-and-error messiness, very well might. "The biggest danger in the human throat's design is choking," writes Nathan Lents. "If we had separate openings for air and food, this would never happen. Swallowing is a good example of the limits of Darwinian evolution. The human throat is simply too complex for a random mutation—the basic mechanism of evolution—to undo its fundamental defects. We have to resign ourselves to the absurdity of taking in air and food through the same pipe."[15]

Abby Hafer, in her pointedly titled book, *The Not-So-Intelligent Designer: Why Evolution Explains the Human Body and Intelligent Design Does Not*, sounds a similar note. "A better designed system would keep the tubes for air and food separate to avoid unnecessary fatalities," she writes. "If we were designed why did the Designer do this job so badly? Or is it that the Creator likes other animals better? There are creatures in which the air passages and food passages are entirely separate. The whale's respiratory system is separate from its digestive system. This means that a whale, unlike a human, can't choke on its food by inhaling it. If the Creator could do that for the whales, I don't know why he couldn't do it for us."[16]

These arguments are riddled with problems. To see why, we need to take a closer look at the human pharynx.

How It Works

In addition to the structures identified in Figure 9.1, fifty different pairs of muscles, controlled by six different nerves, are needed to swallow. After food in the mouth has been formed into a small ball (bolus), the tongue voluntarily moves it to the pharynx, which automatically triggers the involuntary swallow reflex.

As the bolus enters, the pharynx sends sensory information to the swallow center in the brainstem, which immediately turns off respiration so air is not breathed in during swallowing. This prevents the lungs from drawing food into the airway. The brainstem also sends precisely ordered signals telling the various muscles to contract and move the bolus downward into the esophagus, bypassing the airway. This takes about a second.

As swallowing begins, several muscles contract to move the bolus into the pharynx, while moving the back of the palate and the upper pharynx close together to close off the path to the nose.

Next comes the tricky part. The bolus has been blocked from going up into the nose, and muscular contraction is hurtling it down towards the airway and the esophagus. Three separate actions take place to protect the airway. First, muscles contract to close the larynx, which is the gateway to the lungs. Second, other muscles move the larynx up and forward (which you can feel in the front of your neck while swallowing) to hide it under the floor of the mouth and the base of the tongue while being protected by the epiglottis. Third, this action, combined with other muscular activity, opens the upper esophagus to allow the bolus to enter.[17]

The timing and coordination are remarkable. The swallow center must send the right signals via the right nerves to the right muscles, with the exact right split-second timing. Since all this is triggered by the bolus entering the pharynx, the signals from throat to brainstem and back to the many muscles involved (with their reaction times) must be fast enough to prevent choking.

While critics seem to miss the amazing design of this system, it should give the reader pause. Somehow, swallowing happens, usually without incident, a thousand times a day. Where did the information come from that specifies the size, shape, position, and range of movement of the pharynx, each of its nearby structures, and the fifty pairs of muscles involved in swallowing? How could such a system come about gradually, by accident? Where did the information come from to make the swallow center in the brainstem and the logic it uses to control safe swallowing? Where is the repository for the information needed to orchestrate the precisely ordered, well-coordinated contraction sequence of fifty pairs of muscles?

Scoring the Pharynx-Is-Poorly-Designed Argument

With that primer on the pharynx, we are now in a position to see five ways the bad-design argument against it fails.

1. Not Understanding the Design of the Pharynx

The pharynx affords us the dual abilities to breathe and swallow food and water, but it does much more. It affords the ability for speech, language, and tonal activities like lyrical speech and singing. The percussion and acoustic shaping of the tongue, teeth, throat, oral and nasal cavities, and most of the other parts of the pharynx, are required for the nuanced communication that's essential to the human experience. So the pharynx has at least three major functional design objectives. If you were asked to design a system with these capabilities, how would you approach it? How would your design manage all this with a single system? If you used separate systems, as advocated by the critics above, how would you achieve the right kinds of functions, and how would this affect how these functions are packaged into the body as a whole? The critics ignore these questions, apparently because they haven't bothered to understand the design of the system, *as a system*—either its core objectives or the orchestration of its many parts.

2. Not Considering Trade-Offs

Clearly, the pharynx's main three functions cause design conflicts that must be solved. We could use two or maybe even three separate systems to achieve these vastly different goals. However, since all three functions need similar components, two or possibly three copies of many of these structures would be necessary. If, as the critics recommend, we were structured to use the mouth only for swallowing food and water, and not for breathing, thereby precluding speech and language as we know it, the nasal passageways would need to be much larger to bring in enough oxygen during high levels of activity.

To keep all three functions, duplication of parts may be an option. We'd need two mouths, one for eating and another for breathing and speaking, and we'd need two large pipes, one for air and the other for food. We'd need two tongues, one for manipulating food in the eating mouth, and another for speaking in the breathing/speaking mouth. For making the hard consonant sounds in speech, we'd need something like teeth in the breathing/speaking mouth, but we'd also need teeth for chopping up food in the eating mouth. For making complex tonal sounds, the nasal cavities would need to be attached to the breathing/speaking mouth. But we'd also need the nose's smell sensors in the eating mouth in order to fully experience the taste of our food. We could go on, but you get the idea.

In the end, the anatomical changes for either scenario, precluding or preserving speech and language as we know it, would require a complete reconfiguration of the head and neck and possibly also some parts of the lungs and stomach in the body's core. At a minimum, an increase in the size of the nasal passageways would require the head and face to be much wider. But to house duplicate systems, the volume of the head and neck would need to roughly double, and depending on the positioning of the two mouths, the passageways to the lungs and stomach would likely need to be rearranged too.

Maybe if our bodies were shaped more like a whale, this would work better, but of course this might make it harder to climb mountains. Or even to turn our heads quickly.

Building these different functions into a single set of components, with the programming and orchestration to make them work properly, is another example of elegant invention. The obvious trade-off is that it's possible to choke, regardless of how well-designed the system that's in place to avoid this problem.

The marvel is that the system combines these three separate functions in such a compact space, and the whole works so well at all three of its core functions.

3. Not Acknowledging Pharynx Degradation over Time

How and why do humans die from choking? One common cause of swallowing problems is neuromuscular injury or degeneration related to aging or disease. Since swallowing requires precisely orchestrated contractions of many different muscles, any condition that compromises nerve or muscle function can lead to difficulties in swallowing. Common conditions include stroke, Parkinson's disease, and multiple sclerosis (MS), each of which puts the person at risk for aspirating food into their lungs and choking to death. These represent about half of the annual deaths by choking. One could argue that the body's inability to fight off Parkinson's or MS is also a design flaw, but these are also instances of degradation. Complex systems always degrade over time and generations, so it's unrealistic to think this should never happen to the human species if it were well designed.

Another common cause of choking is user abuse. When a healthy adult takes in too large a piece of food, or doesn't chew sufficiently, or a child takes in a foreign object like a small toy, these objects can get stuck in the airway and choking results. One could insist that the design should have been foolproof against such abuses, but this merely takes us back to the question of trade-offs.

To even hope to make the system abuse-proof, the three functions of the pharynx would have to be divided out into two or three separate systems, and we've already seen the problems that attend that strategy. Moreover, no matter how carefully an engineer designs a product, it's always at risk of being misused and, due to wear and tear, its functional capacity will always lessen over time.

4. Jumping from Poor Design of the Pharynx to No Intentional Design

Even if we were to grant for the sake of argument that the pharynx is a case of shoddy engineering, it wouldn't follow that it wasn't intentionally designed, as the Yugo car aptly illustrates. The evolutionists who reach this unsound conclusion perhaps get there by embracing the false premise that poorly designed things must be unintentionally designed things, and combining it with the equally mistaken view that the pharynx is a botched design.

5. Aesthetic Considerations in Evaluating the Pharynx

The irony is that if the designer of the human body had taken their advice and used the vastly clunkier and less elegant approach of creating two or three separate systems for breathing, eating/drinking, and communicating in order to minimize choking, the anti-design critics might have lodged an aesthetic argument against such a choice, namely that no properly ingenious and "tidy-minded engineer" would have failed to elegantly combine the three primary functions into a single clever system.

Engineers know this game—damned if you do and damned if you don't, with critics ignoring the question of trade-offs. Engineers develop thicker skins as a natural coping mechanism. (Which, come to think of it, is another clever adaptive design feature of the human body!)

Ingenious Design

Most people swallow a thousand times a day without incident, all the while breathing in enough air, taking in enough food and water, verbally communicating with nuance, and sometimes even singing. Thus, the rare possibility of choking to death provides little actual evidence of incompetent design. Rather, the human pharynx is more accurately viewed as a clever, elegant solution to a complicated set of competing design objectives, with justifiable choices regarding design trade-offs, within rigid constraints. Further, the solution is profoundly well packaged and even provides a way to equalize the air pressure in the middle ear. This is ingenious design.

The pharynx is but one example of how the poor-design-so-no-designer arguments for mindless evolution fall apart on close inspection. This sort of argument is generally offered as part of a grand evolutionary narrative. You're not supposed to notice that the narrative is built on a panoply of questionable presuppositions, educated and sometimes not-so-educated guesses, cherry-picked data, selective ignorance of key details, and dubious factoids. The grand narrative resembles the proverbial description of a fishing net: *a bunch of holes tied together with string.*[18] The narrative may play well with the Darwinian faithful. It may even convince the uninformed that all is in order with evolutionary theory, so no need to ask any questions. But the attentive will notice that there's more storytelling than science going on here.

To any biologist enamored of poor-design-so-no-designer arguments, we offer the following friendly opportunity and challenge: Let's work together on a serious research proposal to assess the system you see as your most compelling example of poor design in the human body. We will include team members from all the relevant medical and engineering disciplines, perform a rigorous design analysis of the system, in the context of the whole (including the creature's entire life cycle), and jointly publish the results.

The bad-design-so-no-designer argument is a bit like diligently reading and rereading *War and Peace*, looking for grammatical errors, typos, and other mistakes. Of course, you're likely to find examples, whether real or contrived. But in the process, you may have missed the point of the book, and its genius. This analogy, of course, isn't perfect. The human genome contains roughly 150 times more information than *War and Peace*, one of the longest novels in the literary canon. Further, *War and Peace* can't generate its own power, maintain a chemical equilibrium different from its environment, or make copies of itself. But the analogy does make our point: Fixating on error, real or apparent, can lead to overlooking the most important and impressive aspects of a work, whether a novel or an organism such as the human body.

10. Toward a Theory
of Biological Design

When a scientific problem seems impossible to solve,
it may be that a bad assumption—either implicit
or explicit—is constraining our search space.
In these instances, we need to think outside the box.
—Itai Yanai and Martin J. Lercher[1]

D uring the 1984 Super Bowl, Apple Computer aired its
now famous "1984" commercial. In a takeoff of George Orwell's
dystopian novel *1984*, Big Brother is speaking on a large TV screen
viewed by a gray army of bald-headed minions. As he drones on about
pure ideology and unification of thought, an athletic blond woman
wearing a bright white shirt and red shorts runs into the auditorium
wielding a large sledgehammer. There's nothing gray about her. She's
being chased by the Thought Police. Just as Big Brother declares, "We
shall prevail!" the runner flings the sledgehammer at the screen, which
explodes, leaving the audience in shock and presumably destroying
Big Brother's stranglehold on "truth."

In sixty seconds Apple went from being a small, geeky computer
company to a national household name. The commercial captured
the public's imagination and solidified Apple's branding for disrup-
tive forward-thinking and human-centered technology. The original
idea of the ad was that these new, small, and inexpensive comput-
ers would enable individuals to achieve their potential. But there
is a broader point. Sometimes the dominant paradigm needs to be

challenged—particularly when it has been weighed in the balances and found wanting. And dehumanizing.

Modern evolutionary theory is such a paradigm.

Think Different

The causal capabilities of modern evolutionary theory leave much to be desired. Surely there must be a good way to explain the origin of the human body—one that better comports with the evidence. After all, we're here, so we must have gotten here somehow. To get better answers, though, we'll need to do more than just repair the old theory.

To improve on current evolutionary theory, a potential replacement must be able to generate innovations in discrete leaps large enough to overcome the causal hurdles that stand in the way. It also must be able to generate major leaps in body plans fast enough to align with the fossil record's pattern of sudden appearances of nearly all major body innovations. And it must be able to generate the vast array of complex specified information and the precision-defined mechanisms and machines underlying all these capabilities.

In short, any new theory will need to better explain the data, using one or more types of causes known to be capable of such outcomes. As a bonus, it also would be nice if the alternative causal explanation was intuitive and had a simplicity and elegance to rival Darwin's theory.

A Better Way

To this end, a new framework is being developed to apply systems engineering thinking to living organisms—a framework that takes an outlook at least as old as Plato and updates it from a modern systems perspective. It's still early times for this work, but already there is evidence this approach offers more compelling answers to the hard problems of the body's origins. This approach offers not only a better explanation for the origin and properties of living systems, but also a potentially more powerful predictive framework for biological and medical discovery.

In our brief overview of this theory below, we'll mainly employ engineering terminology. The more we learn about biology, the clearer it becomes that organisms are replete with exquisitely engineered solutions to hard problems. Also, the field of engineering addresses the same types of problems organisms must solve to be alive, and indeed, engineering language is uniquely suited for talking about these concepts. In fact, engineering language has become nearly ubiquitous in the biology research literature. In a curious twist, contemporary biologists cannot help but use engineering and design language, even as many of them persist in the view that none of what they're describing was actually engineered.

The Framework

The design view has been widely held for millennia. Both Plato and Moses saw the design in living organisms. But after *The Origin of Species*, evolution by natural selection began to slowly replace design as the dominant view of life's origins.

In the 1980s and 1990s, the modern intelligent design community took shape, drawing on recent scientific discoveries to argue that certain features of the natural world, and especially living systems, are best explained by reference to the work of a designing mind. The outcome has been a solid and increasingly fleshed-out case, based on the best data available, that life was designed.

And now that work is being extended and enriched by a theory of biological design informed by systems engineering. The new framework is conceptually simple. Living organisms are designed (and engineered) to occupy a specific range of environments. Each has the innate, pre-programmed capabilities it needs to adjust and adapt relatively quickly, as needed, within that range. Adjustments can be external, like changing colors, growing longer limbs, or having thicker beaks (how things look, which is relatively easy to observe), or the adjustments can be internal, like switching to an alternate metabolic pathway in response to a different food source (how things work, which is harder to observe). Further, each organism has adaptive resilience, to maintain its life and form over many generations.

On this view, the history of life involves the interplay of four major causal factors:

- *Intentional Acts*: One or more designer-engineers have operated in space-time history to generate new body plans and novel features and, as part of that work, infused life with new information, specialized parts, organizational patterns, and process orchestrations.

- *Internal Adaptations*: Each organism has built-in capabilities for resilience and adaptation. Most of these involve controls to effect regulated change—sensing (typically of both internal and external stimuli), control logic (to decide how to respond), and effectors (to make appropriate internal changes). Such adaptations may be in function, or process, or any other relevant characteristic, including appearance.

- *Design Properties*: Every organism exhibits internal architecture—the patterns by which it's designed, assembled, and operated. Certain architectural principles and design patterns promote change, while others inhibit change. Thus, the architecture of a given organism constrains the ways it may change over time and defines the difficulties and costs associated with those changes.

- *Degradation*: Natural forces (like entropy) work constantly against living systems, pushing them toward equilibrium with the environment (death). Each living organism must actively counter these forces or die. Further, it will be necessary to counter these forces across generations if a species is to last long.

As it turns out, engineers have deep experience with each of these factors. We know what each does and doesn't do. We know how each works. We know the limitations of each. We recognize the artifacts of each, and we know what happens to these artifacts over time.

These causal factors have combined to give us the human body, and the millions of other known species.

While this design-engineering framework has the properties of a good theory, it will nonetheless be controversial in many quarters, since it relies on the intentional acts of a master designer-engineer in the history of life, and such thoughts are not allowed in today's scientistic culture. Why? It's certainly not because the data overwhelmingly support modern evolutionary theory. The evidence seems to support the design-engineering framework much better. Rather, this appears to be mainly a worldview issue. The theory of biological design flies in the face of materialist presuppositions.

Other potential detractors have a theistic framework, but when they do science they restrict themselves to the rule of methodological naturalism, according to which science can only consider purely material explanations in its investigations of the natural world. Intentional actors need not apply. However, to do science properly, surely it is more reasonable to follow the evidence wherever it leads, regardless of prior worldview commitments or methodological dictates. And the theory of biological design is offering more and more evidence that its approach is the better one, as it offers both better answers to longstanding questions, and better questions for future research.

It will take work to turn this into a mature theory. It will undoubtedly be modified and improved along the way. But that's also a strength, since it means many of the findings lie over the horizon, awaiting enterprising and capable scientific investigators to discover them.

Darwinism or Design

We can group all causes into two jointly exhaustive categories. Class 1 causes are purely material causes, lacking foresight and intention. Class 2 causes are intentional causes. The causal hurdles to the origin of the human body—and by extension, to the enormous diversity of living forms on earth, are real and substantial. Intelligent agency (Class 2 causation) regularly surmounts hurdles of this kind in designing and engineering systems. In contrast, Class 1 causal forces, by themselves, have never been shown to overcome such hurdles. While they've long been hypothesized to be able to, such capabilities have never been observed.

So for those claiming that Class 1 causes can engineer life's diversity of organisms, or even just the human body, the burden to demonstrate this falls to them.

Since Darwinian materialists seek to exclude intentional design from a comprehensive explanation for the origin of life's diversity, they would do well to show how purely material processes could gradually build up, over generations, every known biological system and subsystem, including the many examples we've presented in the human body. But lest we come across as piling on, we suggest a more modest challenge for starters: Provide a detailed step-by-step pathway of even just one complex new organ, system, or subsystem discussed in these pages—where each step uses only a small, random change, and with no wishful thinking, no stretches of debilitating dysfunction, and no magic along the way.

We predict that no evolutionist will meet this challenge. Not because the evolutionists lack pluck or intelligence, but for the same reason that even the most brilliant mathematician will never prove that Pi equals two, and the most brilliant engineer will never discover the engineering trick to building a perpetual motion machine. Evolutionists won't uncover any credible, causally adequate evolutionary pathways because none exist.

Short of the kind of demonstration urged above, the Darwinists are asking us to trust that they'll eventually figure it out. This amounts to a promissory note, writ large. But given their inability to deliver a single such pathway even for a short enzyme cascade or complex molecular machine (never mind for a vastly more sophisticated animal system or subsystem), and given that evolutionary theory demands that untold millions of traversable evolutionary pathways exist, it's difficult to grant the Darwinists the fealty that many of them demand for their theory.

Finally, we anticipate that the theory of biological design—outlined in much greater detail in our book *Your Designed Body*—will eventually, in a more mature form, replace Darwinism and its materialistic descendants. Because of the worldview implications, however, this may take a long while.

Conclusion

This is one of the most important intellectual issues of modern times, and every thinking person has the right and duty to judge for himself.
—David Gelernter, "Giving Up Darwin"[1]

You may have been told, often and with force, that you and every other human—indeed, every other living thing—are a cosmic accident. You were also likely told that this is the only reasonable and properly modern view of things. It isn't.

We've shown that there's another option available. The evidence we've presented shows that your body could not have been an accident, that a master designer-engineer must have been involved.

There's the old saying that the captain always goes down with his ship. The saying, hopefully an exaggeration, does carry with it an insight into human psychology. Those who are most invested in a venture are the least likely to abandon it when both prudence and reason urge them to do so. Darwin encountered some serious problems with his theory, but he refused to abandon his ship. In the 160-plus years since he published his theory, the evidential challenges to the theory have grown, including the sort explored in these pages. Committed evolutionists who have invested their careers in the modern form of the theory may also choose to stay with the ship.

But why should the rest of us stay on board as the ship founders?

There are two deals on the table. You are either a cosmic accident, or you are the result of intentional acts by a master designer-engineer. You cannot, of course, be both. You can, however, freely consider both possibilities rather than merely accepting materialism on faith.

So there is one final question to grapple with: What will you do with the information and arguments presented in these pages? Based even on just the evidence for design in the human body, there is sufficient warrant to give the design-engineering framework serious consideration.

We think the evidence is overwhelming, and we are not alone. The famous atheist philosopher Antony Flew did so as well (forcing him to bid adieu to his atheism), as does famous Yale polymath David Gelernter, who in 2019 announced publicly that he had left the Darwinian fold. In fact, recent years have seen a growing number of prominent defections from the Darwinian-materialist viewpoint, and in many cases these people have risked their careers and livelihoods to do so. The number of people who have joined this company continues to grow and now includes scientists, intellectuals, and thought leaders from many backgrounds: Dean Kenyon, Richard Sternberg, Günter Bechly, Thomas Nagel, and many others.

Meanwhile, Nobel Laureate scientists such as Charles Townes and Brian Josephson have explicitly endorsed intelligent design in physics and cosmology, with the latter endorsing two intelligent design books that made ID arguments in both physics and biology—*Return of the God Hypothesis* by Stephen Meyer and *Foresight* by Marcos Eberlin, the latter of which was also endorsed by two other Nobel Laureate scientists. A less widely known trend: Many others have abandoned Darwin but are in sensitive career positions that make it difficult to issue public declarations to this effect.[2]

For many of these who are changing their views, the effort required to maintain belief in the Darwinian status quo simply became too much in light of recent scientific discoveries. This book has explored additional evidence, and when combined, the weight of the evidence is difficult to ignore. So we put the key question to you: How hard are you willing to work to avoid the obvious?

For those who insist that the case for mindless evolution of all life is "overwhelming," the burden is on them to answer the many hard questions we've asked in these pages. If they cannot answer with anything more than hand-waving, storytelling, and appeals to authority, a person should think twice before following them.

In our experience, openly considering the evidence for a designer-engineer can "de-constrain" our thinking from materialist philosophy and open us up to new possibilities, both in our scientific research and in our personal lives. It also offers an escape from the dreary nihilism lurking in the warp and woof of the materialist worldview.

We hope these pages have challenged you to think about these things in new ways. As we conclude, we encourage you to embrace what we consider to be your rightful place in the universe as a real, living person, made for a purpose. But for most of us, getting there is a journey. As you ponder these questions, lean in. Explore the options. Evaluate the trade-offs. Consider the implications and outcomes.

The rest is up to you.

ENDNOTES

INTRODUCTION

1. Max Planck, *Scientific Autobiography and Other Papers*, trans. Frank Gaynor (New York: Philosophical Library, 1949), 33–34.

CHAPTER 1

1. As quoted in Marco Piccolino, "Biological Machines: From Mills to Molecules," *Nature Reviews Molecular Cell Biology* 1 (November 2000): 149–152.

2. Michael Denton, *Evolution: A Theory in Crisis* (Chevy Chase, MD: Adler & Adler, 1986), 249–250. See also his *Evolution: Still a Theory in Crisis* (Seattle: Discovery Institute, 2016), 112 ff.

3. Douglas Axe, "Estimating the Prevalence of Protein Sequences Adopting Functional Enzyme Folds," *Journal of Molecular Biology* 341 (2004): 1295–1315. Several other studies have corroborated Axe's numbers using different methods. See, for example, Sean V. Taylor et al., "Searching Sequence Space for Protein Catalysts," *PNAS USA* 98 (2001): 10596–10601.

4. Elena A. Ponomarenko et al., "The Size of the Human Proteome: The Width and Depth," *International Journal of Analytical Chemistry* (2016), https://doi.org/10.1155/2016/7436849. By 2022, the Human Protein Reference Database (http://www.hprd.org) had cataloged 30,047 different human proteins.

5. Discovery Science, "Molecular Machines—ATP Synthase: The Power Plant of the Cell," *YouTube*, January 21, 2013, video, 3:21, https://youtu.be/XI8m6o0gXDY.

6. The Human Protein Atlas, "The Human Cell," *YouTube*, November 29, 2016, video, 4:07, https://youtu.be/P4gz6DrZOOI.

CHAPTER 2

1. Alexander Tsiaras, "Conception to Birth—Visualized," INK Conference, December 2010, TED, video, 9:21, https://www.ted.com/talks/alexander_tsiaras_conception_to_birth_visualized.

2. Additionally, B cells in the immune system can make changes to their own DNA, so their DNA is not identical to that in the other cells of the body.

CHAPTER 4

1. Ezekiel 37:5 (Revised Standard Version).

CHAPTER 5

1. Isaac Newton, *Opticks: Or, A Treatise of the Reflections, Refractions, Inflexions and Colours of Light. The Second Edition, with Additions* (London: 1718), Book III, Query 28. Available at the Newton Project, August 2006, https://www.newtonproject.ox.ac.uk/view/texts/normalized/NATP00051.

2. This summary borrows from Michael Behe, *Darwin's Black Box: The Biochemical Challenge to Evolution* (New York: The Free Press, 1996) 18–22.

3. Behe, *Darwin's Black Box*, 18–19.

4. Charles Darwin, *On the Origin of Species by Means of Natural Selection* (London: John Murray, 1859), 186.

CHAPTER 6

1. Pratap Sriram Sundar, Chowdhury, and Sagar Kamarthi, "Evaluation of Human Ear Anatomy and Functionality by Axiomatic Design," *Biomimetics* 6, no. 2 (June 2021): 31, https://doi.org/10.3390/biomimetics6020031.

CHAPTER 7

1. Cai Emmons, *His Mother's Son* (Orlando, FL: Harvest Books, 2003), 286.

CHAPTER 8

1. Darwin, *On the Origin of Species*, 187.

2. Richard Dawkins, *The Blind Watchmaker: Why the Evidence of Evolution Reveals a Universe Without Design* (New York: W. W. Norton & Company, 2015), 18.

3. Gerd B. Müller and Stuart A. Newman, "Origination of the Organismal Form: The Forgotten Cause in Evolutionary Theory," in *Origination of Organismal Form: Beyond the Gene in Developmental and Evolutionary Biology*, eds. Gerd B. Müller and Stuart Newman (Cambridge, MA: MIT Press, 2003), 7.

4. Gerd B. Müller, "Homology: The Evolution of Morphological Organization," in *Origination of Organismal Form: Beyond the Gene in Developmental and Evolutionary Biology*, eds. Gerd B. Müller and Stuart Newman (Cambridge, MA: MIT Press, 2003), 51.

5. G. H. Lewes, *Fortnightly Review*, November 1, 1868, 508. Available at http://darwin-online.org.uk/converted/Ancillary/reviews/1868_Lewes_A604.html.

6. Charles Darwin to Joseph Hooker, February 1, 1871, *Darwin Correspondence Project*, Letter no. 7471, University of Cambridge, https://www.darwinproject.ac.uk/letter/?docId=letters/DCP-LETT-7471.xml.

7. For more on neutral evolution, see Stephen C. Meyer, *Darwin's Doubt: The Explosive Origin of Animal Life and the Case for Intelligent Design* (New York: HarperOne, 2013), 321–329, and Michael J. Behe, *Darwin Devolves: The New Science About DNA That Challenges Evolution* (New York: HarperOne, 2019), 94–102.

8. The Royal Society dedicated an issue of their journal to the conference. See Denis Noble et al., "Introduction: New Trends in Evolutionary Biology: Biological, Philosophical and Social Science Perspectives," *Interface Focus* (August 18, 2017), https://royalsocietypublishing.org/toc/rsfs/2017/7/5. See also Paul Nelson and David Klinghoffer, "Scientists Confirm: Darwinism is Broken," *CNSNews*, December 13, 2016, https://www.cnsnews.com/commentary/david-klinghoffer/scientists-confirm-darwinism-broken; Kevin Laland, "Schism and Synthesis at

the Royal Society," *Trends in Ecology and Evolution* 32, no. 5 (May 2017): 316–317; and Susan Mazur, "Pterosaurs Hijack Royal Society Evo Meeting," *Huffington Post*, November 21, 2016, https://www.huffpost.com/entry/pterosaurs-hijack -royal-s_b_13131246.

9. For more on the 2016 Royal Society meeting and other dissent from neo-Darwinism, see Casey Luskin, "*Nature Communications* Retroactively Concedes a Lack of Evidence for Darwinian Gradualism," *Evolution News and Science Today*, March 8, 2022, https://evolutionnews.org/2022/03/nature-communications -retroactively-concedes-a-lack-of-evidence-for-darwinian-gradualism/.

10. James A. Shapiro et al., *The Third Way: Evolution in the Era of Genomics and Epigenomics*, accessed September 12, 2022, https://www.thethirdwayofevolution.com/.

11. Darwin, *On the Origin of Species*, 189. However, even in Darwin's day, sometimes referred to as the Golden Age of Engineering, enough was known to provide him with more than ample cases to abandon his theory. He should have talked to a few of his contemporary engineers.

12. See Meyer's *Darwin's Doubt*, Chapters 15 and 16.

13. Denton, *Evolution: A Theory in Crisis* (1986) and *Evolution: Still a Theory in Crisis* (2016).

CHAPTER 9

1. Thomas Pynchon, *Gravity's Rainbow* (New York: Viking, 1995), 251.

2. Kenneth Miller, *Finding Darwin's God* (New York: Cliff Street Books, 1999), 101.

3. Nathan H. Lents, *Human Errors: A Panorama of Our Glitches, from Pointless Bones to Broken Genes* (New York: Houghton Mifflin Harcourt, 2018), 29.

4. Richard Dawkins, *The Blind Watchmaker* (New York: W.W. Norton, 1986), 93. Tasked with solving a truly hard problem, few engineers would view tidiness as a primary design goal. One wonders how many engineers Dawkins consulted with before making this statement.

5. Charles Darwin to Joseph Hooker, July 13 [1856], *Darwin Correspondence Project*, Letter no. 1924, University of Cambridge, https://www.darwinproject.ac.uk /letter/DCP-LETT-1924.xml.

6. For more on the theological component of Darwinian theory, see Cornelius G. Hunter, *Darwin's God: Evolution and the Problem of Evil* (Eugene, OR: Wipf & Stock, 2019).

7. For an example of this kind of thing, see the embarrassingly naive video on "Stupid Design" from astrophysicist Neil deGrasse Tyson, "Neil DeGrasse Tyson—Stupid Design," *YouTube*, November 8, 2009, video, 4:55, https://www.youtube .com/watch?v=4238NN8HMgQ. He cites several of the examples, including the choking hazard of the pharynx, that we touch on in these pages and in our more in-depth book, *Your Designed Body*.

8. For a book laying out some of the most common bad-design-so-no-designer arguments regarding the human body, see Nathan Lents's 2018 book *Human Errors*. For several articles pointing up the shortcomings of such arguments, and explicitly responding to Lents, search "Lents Human Errors" at *Evolution News and Science Today* (evolutionnews.org).

9. Stephen Jay Gould, *The Structure of Evolutionary Theory* (Cambridge, MA: Belknap Press, 2002), 111–112.

10. Darwin, *On the Origin of Species*, 450.

11. Jonathan Wells, *Zombie Science: More Icons of Evolution* (Seattle, WA: Discovery Institute Press, 2017), 115–124.

12. Ann K. Gauger, Ola Hössjer, and Colin R. Reeves, "Evidence for Human Uniqueness," in *Theistic Evolution: A Scientific, Philosophical, and Theological Critique* (Wheaton, IL: Crossway, 2017), 485–486.

13. Lents, *Human Errors*, 28.

14. For a fascinating discussion of the wrist, see Donald A. Neumann, *Kinesiology of the Musculoskeletal System: Foundations for Rehabilitation*, 2nd ed. (St. Louis, MO: Mosby Elsevier, 2010), 216–243. Especially note Figure 7–15 (227), which shows one dimension of the amazing double hinge of the human wrist, illustrating how two different joints share the same axis of rotation.

15. Lents, *Human Errors*, 19–20.

16. Abby Hafer, *The Not-So-Intelligent Designer: Why Evolution Explains the Human Body and Intelligent Design Does Not* (Eugene, OR: Cascade Books, 2015), 72–73. Others employing the pharynx to plump for evolutionary theory could be cited. Ardea Skybreak, for example, insists that if the human pharynx had been designed, then the designer was "really stupid (or perversely sadistic)." *The Science of Evolution and the Myth of Creationism: Knowing What's Real and Why It Matters* (Chicago: Insight Press, 2006), 109.

17. To see the amazing and beautiful coordination that occurs in swallowing, see "Swallowing Reflex, Phases and Overview of Neural Control, Animation," Alila Medical Media, *YouTube*, April 19, 2014, video, 2:58, https://www.youtube.com/watch?v=YQm5RCz9Pxc.

18. The origin of this joke is unknown—attributed to a "jocular lexicographer."

Chapter 10

1. Itai Yanai and Martin J. Lercher, "What Puzzle Are You In?," *Genome Biology* 23, no. 179 (2022): 5, https://doi.org/10.1186/s13059-022-02748-1.

Conclusion

1. David Gelernter, "Giving Up Darwin: A Fond Farewell to a Brilliant and Beautiful Theory," *Claremont Review of Books*, Spring 2019, https://claremontreviewofbooks.com/giving-up-darwin/.

2. See the documentary *Expelled: No Intelligence Allowed*, directed by Nathan Frankowski, with host Ben Stein (Premise Studios, 2008). Available at *YouTube*, June 14, 2012, https://www.youtube.com/watch?v=V5EPymcWp-g). See also David Klinghoffer, "High-Level Defectors from Evolutionary Theory Leave a Top Darwin Defender Feeling 'Disturbed,'" *Evolution News and Science Today*, August 30, 2012, https://evolutionnews.org/2012/08/high-level_defe/; and John West, "Do Scientists Have the Intellectual Freedom to Challenge Darwinism?," in *The Comprehensive Guide to Science and Faith: Exploring the Ultimate Questions About Life and the Cosmos*, eds. William Dembski, Casey Luskin, and Joseph Holden (Eugene, OR: Harvest House, 2021), 399–406.

FIGURE CREDITS

Figure 1.1. A typical human cell. "Components of Eukaryotic Cell." Image by Christoph Burgstedt, Adobe Stock. Standard license.

Figure 2.1. The three germinal layers of a developing embryo. "Module 9: Human Development and Anatomy Through Lifespan." Image by Rice University, Anatomy 337 eReader, https://creativecommons.org/licenses/by/4.0/. CC-BY-4.0 international license.

Figure 3.1. Fetal and maternal interface. "Placental Structure and Circulation." Image by Sakurra, Adobe Stock. Standard license.

Figure 3.2. Chambers and valves of human heart. "Heart Anatomy." Image by Designua, Adobe Stock. Standard license.

Figure 3.3. Fetal heart. "Module 13: Heart and Great Vessels." Image by Rice University, Anatomy 337 eReader. CC-BY-4.0 international license, https://creativecommons.org/licenses/by/4.0/.

Figure 4.1. Respiratory system. "The Respiratory System." Image by the Canadian Lung Association, 2021. Used with permission.

Figure 5.1. Eye anatomy. "Eye Anatomy with Labeled Structure Scheme." Image by Oleh, Adobe Stock. Standard license.

Figure 5.2. Tissues at the back of human eye. "Eyeball in Section, Structure of Retina." Image by Lavreteva, Adobe Stock. Standard license.

Figure 5.3. Trochlea. "Human Eye Anatomy / Vintage Illustration from Meyers Konversations-Lexikon 1897." Image uploaded by Hein Nouwens, Adobe Stock. Standard license. Modified by Steve Laufmann.

Figure 5.4. Pathway for visual information. "Visual Pathway Medical." Image by VectorMine, Adobe Stock. Standard license.

Figure 6.1. The parts of the ear. "Anatomy of Ear." Image by LuckySoul, Adobe Stock. Standard license.

Figure 6.2. Middle ear. "Middle Ear." Image by Bruce Blaus, October 15, 2013, Wikimedia Commons. CC-BY-3.0 license, https://creativecommons.org/licenses/by/3.0/.

Figure 6.3. Inner ear. "The Internal Ear." Image by Bruce Blaus, October 15, 2013, Wikimedia Commons. Free cultural work license.

Figure 6.4. Cochlea. "Cross Section of the Cochlea." Image by Oarih, March 27, 2010, Wikimedia Commons. CC-BY-SA 3.0 license. Modified by Steve Laufmann.

Figures 8.1a and b. Partial problem cascade for cellular respiration. Image by Steve Laufmann.

Figure 9.1. Pharynx. "Head and Neck Overview." Image by Arcadian, February 15, 2007, Wikimedia Commons. Public domain.

Steve Laufmann is a speaker, author, computer scientist, and consultant in the design of enterprise-class systems, with expertise in the difficulties of changing complex systems to perform new tasks. Co-author of the 2022 book *Your Designed Body*, he leads the Engineering Research Group at Discovery Institute and chaired the program committee for the 2023 Conference on Engineering in Living Systems.

Dr. Howard Glicksman is a general practitioner with more than forty years of medical experience in office and hospital settings, who now serves as a hospice physician seeing terminally ill patients in their homes. He received his MD from the University of Toronto and is the author of "The Designed Body" series for *Evolution News and Science Today* and co-author of *Your Designed Body* (2022).

Watch the Video Series

SecretsoftheHumanBody.com

Continue to Explore

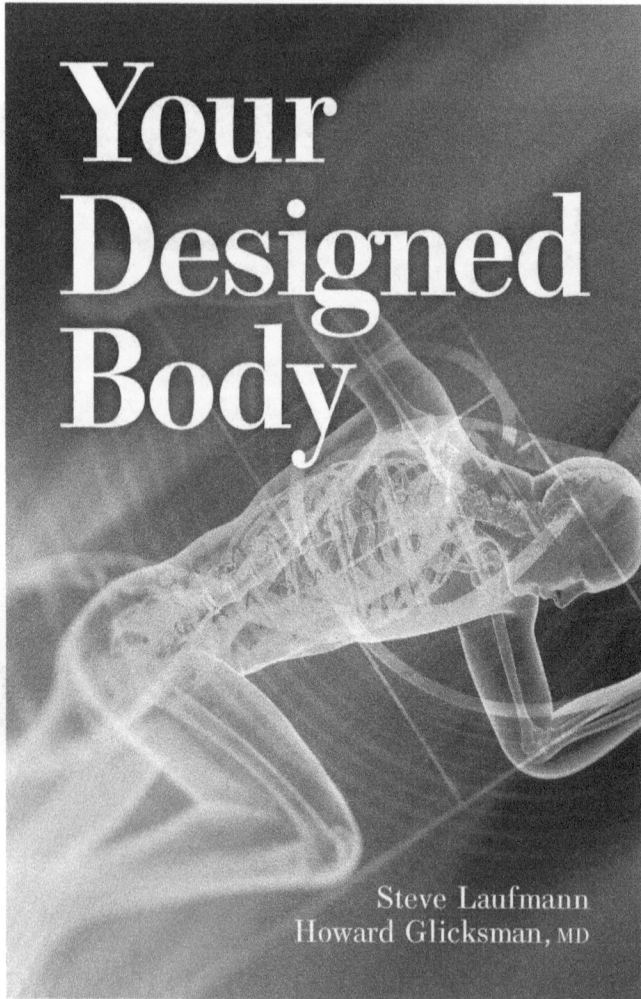